Feeding dairy cows in the tropics

Edited by
Andrew Speedy
and
René Sansoucy

Reprint by arrangement with the Food and Agriculture Organization of the United Nations by Daya Publishing House®
A Division of
Astral International Pvt. Ltd.
New Delhi - 110 002

"This book was originally published by the Food and Agriculture Organization of the United Nations (FAO) as Feeding dairy cows in the tropics, FAO Animal Production and Health Paper, 86".

"The designations employed and the presentation of material in this publication do not imply the expression of any opinion, whatsoever on the part of the Food and Agriculture Organization of the United Nations concerning the legal status of any country, territory, city or area or of its authorities, or concerning the delimitations of its frontiers or boundaries."

'The designations 'developed' and 'developing' economies are intended for statistical convenience and do not necessarily express a judgment about the stage reached by a particular country, territory or area in the development process.'

First Indian Edition, 1998
Reprinted, 2016
ISBN: 978-81-7035-210-5

Published by : **Daya Publishing House®**
A Division of
Astral International Pvt. Ltd.
– ISO 9001:2008 Certified Company –
4760-61/23, Ansari Road, Darya Ganj
New Delhi-110 002
Ph. 011-43549197, 23278134
E-mail: info@astraiint.com
Website: www.astralint.com

Printed at : **Chawla Offset Printers,**
Delhi-110052

CONTENTS

(*) These authors were unable to participate in the meeting. However their paper is included as a valuable contribution to these proceedings.

INTRODUCTION

The FAO Expert Consultation on Feeding Dairy Cows in the Tropics was held in Thailand in the FAO Regional Office for Asia and the Pacific in Bangkok, from 3 to 7 July 1989.

Recent FAO statistics show that, while milking the same number of cows (about 110 million head) the developing countries (mainly located in the tropical zone) produce only 22 % of the whole fresh milk equivalent produced by the developed countries and 18 % of the total world production (461.5 million t). In addition, milk production in Asia and to a lesser extent in Africa was reduced from 1986 to 1987 due to drought and the policy measures taken by some countries. In spite of successful achievements such as "Operation Flood" in India, many failures have also been observed in the past.

The problems encountered in stimulating milk production in developing countries are very complex. As in other agricultural development operations many difficulties such as pricing, marketing, etc are beyond the control of the producer. However, technical constraints including nutrition, health and breeding, have still to be, and can be overcome.

Among these constraints the nutrition aspect is probably the first factor limiting milk production. The 1987 Conference of FAO drew attention to the increasing difficulty in providing the bulk of feed requirements for cattle through grazing, crop by-products and to a lesser extent, fodder crops.

Recent advances in the knowledge of ruminant nutrition physiology and in the nutritive value and techniques of utilization of feed resources including unconventional ones e.g. crop residues and agro-industrial by-products, provide scope for overcoming the forecasted feed shortage.

The purpose of this expert consultation therefore was to:

- review the various milk production systems in the tropical areas (humid and dry) according to agro-climatic and technical, economical and sociological conditions including special situations like peri-urban production systems;

- review new knowledge in ruminant digestion nutrition and physiology and consider ways and means of implementing rational feeding systems that could overcome, at the lowest cost, nutritional constraints which hamper milk production and herd productivity in the various prevailing systems;

- match milk production (specialized or dual purpose) systems to available and potential feed resources, taking into account their nutritional characteristics, and considering both subsistence and commercial systems;

- make recommendations for the development of sustainable milk production systems based on locally available feed resources

The opening speech was delivered by Mr. Vitoon Kamnirdpeth, Director General of the Department of Livestock Development, on behalf of the Royal Government of Thailand. First of all he expressed his government's appreciation to FAO for organizing this Expert Consultation in Thailand, and welcomed all the participants. He briefly reminded the Experts of the milk production history in Thailand which started only 30 years ago and has developed rapidly during the last ten years. Although milk production is increasing, milk consumption per caput is still low and the Government is trying to encourage the Thai population to consume more milk and milk products. He pointed out that the main problems that the dairy farmers encounter in Thailand are similar to those encountered in other tropical developing countries, particularly those concerning animal feeding but also breeding, health, etc... This Expert Consultation will deal with the feeding of dairy cattle. It should be a good opportunity within this context to help increase milk production in tropical developing countries.

The welcome address was given by Mr. S.S. Puri, Assistant Director-General and FAO Representative for Asia and the Pacific. He first pointed out that the Expert Consultation would not only deal with dairy cows but also with buffaloes which are very important in Asia, especially in India where they are more numerous than cows. He mentioned that in Asia human diet is mainly dependent on rice/cereals which implies an unbalanced diet. Thus, in terms of non-cereal food diet, there is a need for increasing milk and meat production and availability. About 40 % of the number of cows in the world are in Asia, but they are only responsible for about 7 % of the world milk production. The productivity of the animals is low and did not follow the rapid increase of cereal yield in the last 20 years. For these reasons, diets are still deficient in good quality proteins and particularly those from milk and milk products. He also pointed out that in the tropical areas, there is very little land still available for pastures. For example, in India 96 % of the usable land is already cultivated. There is a need to increase feed in quantity and quality for milk production. Therefore this Expert Consultation on feeding is of great importance.

Technical secretaries were R. Sansoucy and P. Hassoun

LIST OF PARTICIPANTS

AUSTRALIA:	R.A. LENG Professor of Nutritional Biochemistry Department of Biochemistry, Microbiology and Nutrition University of New England Armidale, N.S.W. 2351 Telex: 166050 Tel: (067) 732707 Telefax: (067) 733122 E-mail: rleng@gara.une.oz.au (UUNET network)
CHINA:	CHEN NANQING Senior Researcher in Tropical Dairy Production Institute of Animal Husbandry Academy of Agricultural Sciences Guangdong c/o FAOR: Telex: 22664 FAO BJCN (CHINA) Tel: 5322835/6/7 PABX
CUBA:	R. GARCIA TRUJILLO Sub Director Instituto de Ciencia Animal Apartado 24, San José de las Lajas La Habana c/o FAOR: Telex: 512110 FAO CU (CUBA) Tel: 219717, 219155, 218470 (direct)
EGYPT:	A.M. EL SERAFY Professor of Animal Nutrition Ain Shams University 12 Kobry El-Kobba St. Cairo c/o FAOR: Telex: 21360 FOODAG UN Tel: 709084 (direct), 705182, 705029, 703412, 703475 Ext: 156
FEDERAL REPUBLIC OF GERMANY:	K.H. LOTTHAMMER Professor, Animal Health Office Chamber of Agriculture Weser-Ems Mars-La-Tour Strasse 1 2900 Oldenburg Telex: 25639 (Lakaol) Tel: 0441/801-640 Telefax: 0441/801180

FRANCE:

M. CHENOST
Senior Research Officer
Institut National de la Recherche
Agronomique
63122 Theix, Ceyrat
Telex: 990227F (INRATEX)
Tel: 73624075
Telefax: 73624450

Y. CHILLIARD
Senior Research Officer
Institut National de la Recherche
Agronomique
63122 Theix, Ceyrat
Telex: 990227F (INRATEX)
Tel: 73624000
Telefax: 73624450

INDIA:

V. BADVE
Senior Research Officer
Bhartiya Agro-Industries Foundation
Atx Post Urulikanchan - 412202
Dist. Pune India
Telex: 145283 BAIF IN

MALAYSIA:

Aminah ABDULLAH
Senior Research Officer
Livestock Research Division,
Malaysian Agricultural Research and
Development Institute (MARDI)
P.O. Box 12301, General Post Office,
50774 Kuala Lumpur
Telex: MARDI MA 37115 VVV
Tel: 03-9486601 Ext. 435

MALI:

M. TRAORE
Director
Centre National de Recherches
Zootechniques
B.P. 262, Bamako
Telex: 2426 FOODAGRI BAMAKO
Tel: 355055

MAURITIUS:

A. BOODOO
Senior Researcher
Animal Production Division
Ministry of Agriculture
Le Reduit
c/o UNDP:
Telex: 4259 UNDP IW (MAURITIUS)
Tel: 23726/7
Telefax: (203) 084871 UNDP PORTLOUIS

PHILIPPINES:	F.A. MOOG Chief Agricultural Development Specialist, Research Division Bureau of Animal Industry (BAI) Ministry of Agriculture Alabang - Metro - Manila c/o FAOR: Telex: 64111 FAO PN or 23149 FAOR PH Tel: 850611/30 Ext:259, 8171654 (dir)
TANZANIA:	L. MORUNGU National Team Leader FAO Project URT/86/013 Assistance to Smallholder in Dairy Development P.O. Box 3092 Moshi c/o FAOR: Telex: 41320 FOODAGRI/KADC 43005-TZ (MOSHI) Tel: 32979 (direct), 33914, 33917, 33910
THAILAND:	R.W. FROEMERT Team leader FAO Regional Dairy Development and Training Team for Asia and the Pacific P.O. Box 125 Chiang Mai University Chiang Mai 50002 Thailand
	S. KHAJARERN Professor Department of Animal Science Faculty of Agriculture Khon Kaen University Khon Kaen 40002 Telex: 55303 UNIKHON TH
	S. PICHET Dairy Farming Promotion Organization of Thailand Muaklek, Saraburi Tel: (036) 244167

UNITED KINGDOM:

T.R. PRESTON
Scientific Adviser
Fundación para el Desarrollo
Integral del Valle del Cauca
Programa CIPAV
Camera de Comercio
Apartado Aereo 7482
Cali, Colombia
Telex: 055724
Tel: 823271
Telefax: 824627 CALI COLOMBIA

UNITED STATES OF AMERICA:

H.D. JOHNSON
University of Missouri-Colombia
Dairy Science Department
114 Animal Science Research Center
Colombia, Mo.
65211 USA
Telex: 704786 UM INTL AG UD

FAO:

P. HASSOUN
Associate Professional Officer
(Feed Resources), Animal Production
and Health Division FAO HQ
via delle Terme di Caracalla
00100 Rome
Telex: 610181 FAO I
Tel: 57971
Telefax: 57973152/5782610

R. SANSOUCY
Senior Officer
(Feed Resources), Animal Production
and Health Division FAO HQ
via delle Terme di Caracalla
00100 Rome
Telex: 610181 FAO I
Tel: 57971
Telefax: 57973152/5782610
E-mail: AGAP01@IRMFAO01
(EARN network)

MEDIUM-TERM OUTLOOK FOR DAIRYING IN THE DEVELOPING COUNTRIES

by

W. Krostitz

INTERNATIONAL BACKGROUND

Milk production and consumption have so far been concentrated in the developed regions - mainly Europe, the USSR, North America and Australasia - though more recently Japan has also become an important milk consuming and producing country. In contrast, the developing countries, with about three quarters of world population, account for just one quarter of world milk output and, as net imports have risen, for slightly over a quarter of global consumption of milk and milk products. However, since the 1970s, when their share was only one-fifth, milk production in the developing countries has grown faster than in the developed regions.

This difference in production and consumption trends between the two categories of countries is likely to accentuate in the medium term. As indicated in Table 1, the developing countries' share of global milk production could reach nearly a third by the end of the century. Yet, as their population is projected to increase to 4.8 billion, or four-fifths of the world total, while their net imports of dairy products will probably decrease, average availability, at just over 40 kg of milk equivalent, will not be significantly higher than at present and will be still merely one-seventh of the average level for the developed regions. Moreover, this relatively low statistical average will conceal wide differences between individual countries and, within countries, among income groups. Perhaps even more than in the past, consumption of milk and milk products in the developing regions will be concentrated among middle and high-income groups in urban areas. However, dairy development could increasingly support efforts to improve rural incomes and living standards in developing countries with a reasonable potential for milk production.

The difference in likely future rates of growth in milk production between developing and developed countries reflects not only trends in demand but also recent changes in production policies. In the developed market economies, with the major exception of Japan, *per caput* demand for milk and milk products overall seems to have reached saturation point, while population growth is slow. In some Eastern European countries and the USSR there would appear to be scope for rises in demand in the long run, but the medium-term outlook is for stagnating or even falling consumption owing to the effects on consumer prices and incomes of structural adjustment programmes. Moreover, with non-food use, particularly feed use, decreasing average availability per head of population in the developed countries as a whole is expected to fall by the end of the century.

Large-scale, heavily subsidized use of milk and milk products in livestock feeding was one of the main features of surplus disposal in the developed market economies over the past quarter century when milk output increasingly exceeded effective demand. On average during the 1980s, governments of West European and North American countries subsidized the feed use of some 20 million tons of milk equivalent per year, most of it in the form of skim milk and skim milk powder but occasionally also butter, with the EEC accounting for the bulk of this. At the same time, West Europe and North America exported large quantities of milk products, at reduced prices, or as food aid. Thus, by the first half of the 1980s, almost 20 million tons of milk equivalent annually or three-quarters of world exports of dairy products were heavily subsidized. Confronted by this situation, low-cost producers of traditional exporting countries in Australasia were hardly able to maintain the volume of their sales abroad, despite a growing international market, while most suppliers among developing countries had to withdraw from export marketing. At the same time, net imports of the developing countries, of which nearly 20 percent was food aid, reached about one-eighth of their total consumption as can be seen from the table.

Table 1. WORLD DAIRY SITUATION AT A GLANCE - PAST AND PROJECTED

	Milk production	Net imports 1	Availability [1 2] Total	Per caput
	(--------million tons		-----------)	kg
World total				
1974-76	422	-	422	106
1986-88	521	-	521	104
2000	585	-	585	95
Developed countries				
1974-76	335	-8	327	292
1986-88	388	-20	368	300
2000	400	-15	385	290
Developing countries				
1974-76	87	8	95	32
1986-88	136	20	156	41
2000	185	15	200	42

[1] Milk and milk products in milk equivalent.
[2] Including non-food use and waste.

Ample availabilities of relatively cheap, or entirely free, milk products in international markets supported policies aimed at low consumer prices in many developing countries which in turn discouraged the development of local milk production. As a matter of fact, in a situation of almost chronically depressed prices in international trade, the comparative advantage of milk production in developing countries was generally considered to be low. This was true not only of the lending policies of national and multilateral development banks but also of the investment policies of transnational companies. Although such companies made considerable investments in milk processing and distribution throughout the developing regions, their interest in the development of local milk production lessened in view of the cheap raw material supplies in international markets that could be drawn upon for recombining. Among the few developing countries which, notwithstanding this discouraging background, launched policies of strong support to domestic dairying as early as in the 1960s and 1970s were India, the Republic of Korea, China, Cuba and, to a certain extent, Venezuela. As in India, dairy development programmes in Cuba and China have benefited substantially from food aid in the form of milk powder and butter oil which after recombining have been sold in the urban markets at prices equivalent to those obtained by domestic producers. While adding to local milk supplies, the main purpose of such commodity aid has been to generate funds for investment into the development of local milk production and marketing.

WORLD STOCKS AND EEC SURPLUSES

However, during the second half of the 1980s, the international dairy situation changed significantly. The EEC, faced with ever rising costs of its farm support policy, introduced quotas on the marketing of cow's milk in 1984/85 and has since then reduced output by about 10 million tons, or one tenth. Other West European countries and Canada have operated similar supply management schemes. The United States has repeatedly cut the milk support price, accompanied by direct measures to reduce the dairy cattle population. In addition, dairying in New Zealand was affected by adverse weather. Helped by huge sales of butter at cut-rate prices mainly to the USSR, world stocks of dairy products were reduced to rather low levels by 1989, while prices in international trade in dairy products rose sharply. On the basis of c.i.f. prices of skim milk powder and butter oil prevailing during the first half of 1989, the cost of raw material for the recombining of 1 tonne of milk with 3 percent fat was about US$260, almost three times as much as during the mid-1980s. At this level of international prices the competitive advantage of imports over local dairy products had practically disappeared in countries with a reasonable potential for dairying. As discussed in a second paper submitted by FAO's Commodities and Trade Division to this seminar, prices in international dairy products trade have decreased again in 1989/90 though they remain well above the levels registered in the middle of the 1980s.

Milk production in Western Europe and, to a lesser extent, North America still exceeds effective demand. Hence, access to their markets will remain restricted and some subsidization of their domestic consumption and exports is likely to continue. However, unless present production controls in these two regions are significantly relaxed surpluses should remain manageable and prices in international dairy trade would not return to the very depressed levels of the mid-1980s. [1]

EFFECTS OF HIGHER INTERNATIONAL PRICES ON DEVELOPING COUNTRY IMPORTS

In the USSR, Eastern Europe and the developing countries which account for the bulk of world imports of dairy products, self-sufficiency ratios will probably increase in the medium term. Higher international prices and continued foreign exchange constraints seem to have discouraged imports by the USSR and some East European countries and encouraged exports by others. For similar reasons, many developing countries have reduced imports, while several low cost producing developing countries may expand or resume exports of dairy products.

In Argentina, Uruguay and southern Chile milk production costs, at about US$100 per tonne, are among the lowest in the world. Although shipments are small compared with those of the big suppliers in Europe, North America and Oceania, Uruguay has in recent years been the largest net exporter of dairy products among the developing countries. Given supportive government policy, Argentina could regain its position as the leading exporter of dairy products among developing countries. Once a significant supplier of butter, cheese and preserved milk products, Argentina's dairy industry has reduced its sales abroad over the past two decades, when the fall in export returns was compounded by taxes imposed by government on shipments to foreign markets. Elsewhere in Latin America, Costa Rica has recently begun to offer milk powder in external markets as rising prices have stimulated milk output and, reinforced by recession, curbed domestic demand for milk and milk products. In the longer run, Nicaragua might become an exporter of milk powder again, having been a net importer of dairy products throughout the 1980s.

In Africa, Zimbabwe and Kenya are re-emerging in their traditional role as net exporters, mainly of milk powder and butter to

[1] Assuming continued restrictive production policies in West Europe and North America, dairy farmers in Oceania will probably feel encouraged to raise production for export though this is not likely to offset the reduction in West European and North American supplies. Japan, the only OECD country whose imports of dairy products could grow over the medium term, might absorb part of higher Oceanian supplies.

neighbouring countries in southern Africa and the Near East. In the markets of the Near East oil countries, Turkey also hopes to be able to sell rising quantities of cheese and other milk products, and there could be interest in certain milk specialities from South Asia. China and Mongolia might increase sales to the eastern parts of the USSR, a number of developing countries in East and South East Asia and perhaps to Japan. [2]

FUTURE PROSPECTS

In the short term the response of existing and potential suppliers to higher international prices, both in the developing regions and in Oceania and East Europe, could be through reducing domestic consumption rather than raising production. The long depression in international prices has discouraged investment into dairy farming in many low-cost producing countries, reinforced by general economic and financial difficulties including high interest rates and shortage of foreign exchange. Because of the large amounts of capital and time required to raise milk production and processing, future prospects for dairy products exports are apparently assessed with caution.

In fact, the largest part of the projected increase in milk production of the developing countries is likely to be for domestic consumption and, as a group, the developing countries will probably continue to have a sizeable net import, as indicated in the table. Imports will be increasingly concentrated in the petroleum exporting countries of Asia, North Africa and Latin America but, unless oil prices rise, even these countries may not maintain the volume of their imports at recent levels. Moreover, some of them, such as Saudi Arabia, have for some time promoted local milk production though in such areas dairying is costly, and a country's dependence on imported dairy products may just be replaced by dependence on feeds and other inputs. Dairying based on imported feeds will therefore be the exception rather than the rule in a situation where many developing countries are heavily indebted and short of foreign exchange.

DEMAND AND PRODUCTION IN ASIA

One country which has been able to afford a rapid expansion not only of its milk but also of its meat and egg production largely based

---

[2]In the longer term a general liberalisation of agricultural markets and trade as a result of the GATT's Uruguay Round would of course greatly improve the prospects for exports from low-cost producing countries such as Uruguay, Argentina, New Zealand or Australia. However, at this stage it would appear to be premature to talk about possible implications of any agreements reached under the Uruguay Round.

on imported feeds is the Republic of Korea. In recent years this country has purchased between 7 and 8 million tons annually of feed grains and other concentrate feeds at a cost of roughly US$ one billion. In addition, the Republic of Korea where milk production was virtually unknown until the 1960s but reached 0.5 million tons in 1980 and over one million tons in 1985, based its dairy development programme essentially on imported dairy cattle. The Korea Rural Economic Institute projects milk production to rise to 1.5 million tons by the early 1990s and nearly 3 million tons by the beginning of the next century. While expensive in terms of foreign exchange and to final consumers, milk production has been profitable to farmers who have specialized in this activity or added dairying to their enterprise. The Republic of Korea, though not providing a model of dairy development which would be suitable for the majority of developing countries, illustrates the change in food consumption habits as a result of rapid economic growth and urbanization. Unlike most other developing regions, East Asia has no tradition of milk consumption. If the above projections materialize, per caput consumption of milk and milk products in the Republic of Korea which was virtually nil before 1970, would reach over 30 kg of milk equivalent in the early 1990s and over 50 kg by the turn of the century. However, notwithstanding its sizeable balance of trade surplus, the country is becoming preoccupied about its dependence on imported feed and is now paying greater attention to the development of domestic fodder resources.

China is another example of rapidly rising demand for and production of milk as a result of social and economic change. Traditionally the majority of Chinese did not consume milk, but since the late 1970s demand for milk and milk products has experienced a fast rise in urban areas around which modern dairying based on high yielding Chinese Black and White cattle (largely Holstein-Friesian blood) and, to a lesser extent goats, is being promoted. By 1992 Chinese plans call for an increase in per caput consumption of milk from 3 to 5 kg. With most milk consumed by the urban fourth of the population, average intake in urban areas would thus reach some 20 kg at the beginning of the next decade. Milk production in China has risen at double-digit rates during the past decade but was still less than 4 million tons in the late 1980s. Original plans which indicated a target of 30 million tons by the year 2000 have been revised downwards as feed supply is lagging behind the requirements of the livestock sector.

In South East Asia imported milk products became popular among urban consumers after the Second World War, but local milk production is a more recent phenomenon, with Indonesia and Thailand experiencing the most rapid growth. As in China, dairy development in these two countries is based on domestic feed and cattle of European or North American origin and crossbreeds. However, whereas China shifted emphasis from dairying on large-scale state or collective farms to milk production by smallholders in the 1980s, Thailand and Indonesia have right from the start given priority to smallholder dairying within their rural development policies.

The same has been true of India, the biggest milk producer among developing countries, where milk output is projected to rise by some 40 percent to 61 million tons by 1995 with per caput consumption increasing from its present level of 58 kg per year to about 68 kg. Under the project "Operation Flood", the world's largest dairy development project, a resource-conscious policy of production largely based on local feeds and indigenous cattle and buffaloes and accompanied by efficient marketing of milk and milk products has benefited both rural producers, especially small- holders, and urban consumers. The success of this project, not only from a technical but also from a socio-economic viewpoint, has in the meantime aroused increasing interest in dairy and agricultural development planning throughout the developing countries.

As in South Asia, milk production and consumption has a long tradition in West Asia and most parts of Africa and Latin America. However, in the latter three regions prospects for dairy development appear to be less favourable than in South and East Asia. This reflects not so much natural resource endowment but rather the outlook for general economic development. During the remainder of the century, economic growth and rises in incomes are likely to be considerably higher in the eastern and southern parts of Asia than in the other developing regions. Hence, as in the recent past, strong consumer demand is likely to be a major stimulus to dairying in southern and eastern Asia, an area which will comprise almost half of the world's population by the turn of the century.

THE FUTURE FOR MILK IN DEVELOPING COUNTRIES

However, although milk production is a relatively efficient way of converting vegetable material into animal food and dairy cows, buffaloes, goat and sheep can eat fodder and crop by-products which are not eaten by humans, the loss of nutrients involved in production and the large amounts, often imported, of energy and equipment required in milk handling inevitably make milk a comparatively expensive food. Also, if dairying is to play its part in rural development policies, the price to milk producers has to be remunerative. In a situation of increased international prices, low availabilities of food aid and foreign exchange constraints, large-scale subsidisation of milk consumption will be difficult in the majority of developing countries. Hence, in the foreseeable future, milk and milk products in the majority of developing countries will not play the same role in nutrition as in the affluent societies of developed countries. Effective demand will come mainly from middle and high income consumers in urban areas.

There are of course ways to mitigate the effects of unequal distribution of incomes. In Cuba where the government attaches high priority to milk in its food and nutrition policy, all pre-school children, urban and rural alike, receive a daily ration of almost a litre of milk at a reduced price, and cheap milk and milk products are made available to certain other vulnerable groups, while milk products outside the rationing system are sold at a price which is well above cost level. Until recently most fresh milk in the big cities of China was reserved for infants and hospitals, but with the increase in supply rationing has been relaxed. In other countries dairy industries have attempted to reach lower income consumers by variation of compositional quality or packaging and distribution methods or blending milk and vegetable ingredients in formula foods for vulnerable groups. For instance, pricing of products rich in butter fat or in more luxury packaging above cost level so as to enable sales of high protein milk products at a somewhat reduced price has been widely practised in developing countries.

THE LACTATING COW IN THE VARIOUS ECOSYSTEMS: ENVIRONMENTAL EFFECTS ON ITS PRODUCTIVITY

by

H.D. Johnson

CLIMATES OF THE WORLD

Milk yields are a product of animal genetic and environmental interactions. Milk yield for a specific genotype, especially in tropical environments or ecosystems, is a function of climate and its interactive influences on the quantity and quality of feed, the presence of disease and parasites and the utilization of technology to alleviate nutritional, thermal and health limitations. Each climate zone and/or ecosystem includes variations in the environmental complex which influences milk yields and dairying practices in the tropics. Zones of the world between the Tropics of Cancer and Capricorn include the majority of the cattle and buffalo of the world and the climate in these regions is especially limiting to milk yields, growth and reproduction when both the temperature and humidity is high.

Emphasis in this paper will be on the thermal or direct climatic aspects of the environmental factors as they influence the cows ability to eat, maintain heat balance, produce milk and reproduce. Of the meteorological factors (temperature, humidity, wind, radiation, photoperiod and rainfall), the temperature and humidity (including rainfall) are the limiting factors most difficult to alleviate.

Figure 1 presents the average monthly temperature-humidity-index (THI) for numerous temperate zones and tropical countries. Generally, the climate of the countries in the left section, depending on the numbers of months above 72 THI, does not preclude the use of Holstein dairy cows. However, the introduction of Holsteins into the countries on the right section (humid tropics) results in moderate to severe limitations in milk yield due largely to the temperature/humidity and related nutritional factors. Adaptable but lower yielding indigenous cattle have been used for centuries as a source of meat, milk and fibre in the tropical zones.

Figure 2 is an illustration of these calculations showing the number of months throughout the year that THIs were greater than 72 (B). THI average of months above 72 (C), which is a multiple of A x B, best expresses the comfort or adversity of climate zones. The annual THI average also reflects the relative limitation of the various climate zones of the world for dairy cows (Figure 3). Table 1 lists these indices which express the adversity or comfort as indicated by Columns A, B, C and D.

Figure 1.

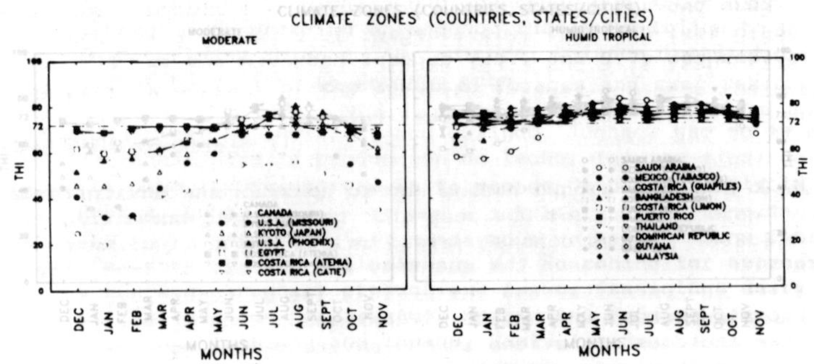

Figure 2.

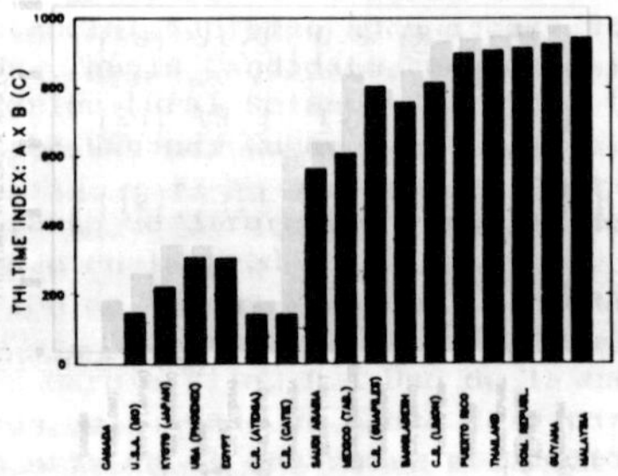

Figure 3.

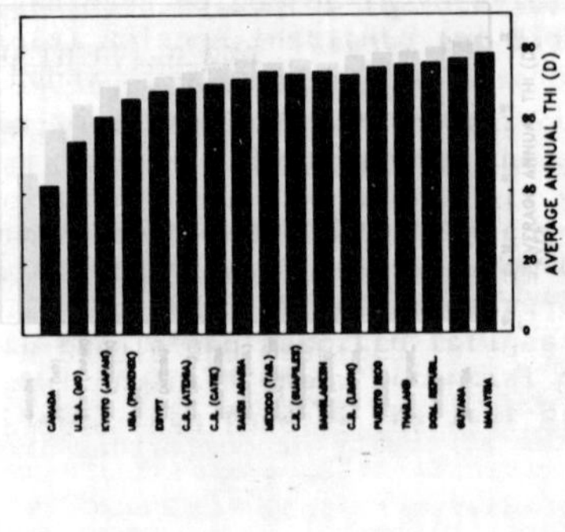

Table 1. Temperature Humidity Index - THI.

THI = T dry bulb + (0.36 T dew point) + 41.2°C

	Climate Zone	A. Ave. months above 72	B. No. months above 72	C. Time index: A x B	D. Average Annually
1	Canada (Edmonton)	0			42.2
2	U.S.A. (Missouri)	74.5	2	149	54.6
3	Japan (Kyoto)	74.7	3	224	61.6
4	U.S.A. (Phoenix)	77.0	4	308	66.6
5	Egypt (So. Delta)	76.5	4	306	68.8
6	Costa Rica (Atenia) (low-land-dry)	71.9	2	144	69.5
7	Costa Rica (CATIE) (mid-altitude-humid)	71.4	2	144	70.7
8	Saudi Arabia (Hufuf)	80.5	7	563	71.9
9	Mexico (Cardenas, Tabasco)	76.0	8	608	74.0
10	Costa Rica (Guapiles) (low-humid)	73.2	11	805	73.2
11	Bangladesh (Dhaka)	75.8	10	758	73.9
12	Costa Rica (Limon) (low-humid)	74.2	11	816	73.0
13	Puerto Rico (San Juan)	75.0	12	900	75.0
14	Thailand (Bangkok)	75.7	12	908	75.8
15	Dominican Republic (Santiago)	76.2	12	915	76.2
16	South America (Guyana)	77.2	12	926	77.3
17	Malaysia (Kuala Lumpar)	78.7	12	944	78.7

THE ENVIRONMENT, THERMAL BALANCE, FEED INTAKE AND MILK YIELD

A model of major factors of the ecosystem (not including other animals) which influence the ability of the cow or other livestock to lactate, grow and reproduce has been described by Johnson, 1987b. The meteorological factors include temperature and photoperiod and involve physiological mechanisms; the most important non-meteorological factors are quantity and quality of feedstuffs and disease factors. Environmental temperature (thermal factors) and possibly emotional factors signal the hypothalamus and central nervous system to alter feed intake, hormonal functions and heat production and/or loss with resultant declines in milk yield and fertility.

In a thermal environment in which the animal's heat production exceeds heat loss, an increasing amount of heat is stored in the animal's body, resulting in increased body temperature. When the body temperature is significantly elevated, a myriad of homeothermic events are initiated. These events include increases in evaporative heat loss by respiration and skin. However, when high temperatures and radiation lessen the ability of the animal to radiate heat from the body, feed intake, metabolism, body weight and milk yields decrease to help alleviate the heat imbalance (Johnson, 1980a,b). Even though tissue substrates are mobilized, energy metabolism, growth and lactation declines.

To avoid this excessive acclimatization or adjustment to an adverse environment or to alleviate the stressor effects on less adaptable individuals, various management decisions and practices may be used to alleviate the severity of the climatic influences on the animal. These practices can help maintain the efficiency of production and prevent disintegration of the animal system.

The level of feed intake as indicated in Figure 4, is determined partially by the thermal balance of the animal which in turn alters milk yields and reproductive performance. Feed or hay intake declines in relation to THI which is illustrated clearly in Figure 4. The decline is about 0.23 kg/day for each unit increase in THI or increase in rectal temperature. The related decline in milk yield with increasing THIs is approximately 0.26 kg/day, milk decline/unit increase in THI (Johnson et al., 1961, 1962; Johnson, 1987b). A more recent study with 52 cows at each stage of lactation demonstrated the relative time changes in rectal temperature and milk yield and feed intake (Johnson et al., 1988). Milk yields of Jerseys and Holsteins from some of the countries previously discussed (Table 1) have been affected by the total environmental complex. These declines in milk yields for Holsteins or Jerseys in a temperate climate as compared to the tropics are very great (approximately half of genetic potential). The somewhat lesser decline in Puerto Rico (half of genetic potential) may be due to data from the "hills" region, with improved genotypes and management practices (Table 2).

Figure 4. Regressions of milk yield, rectal temperature and feed intake on THI for many temperature-humidity conditions above thermoneutral (Johnson et al., 1962).

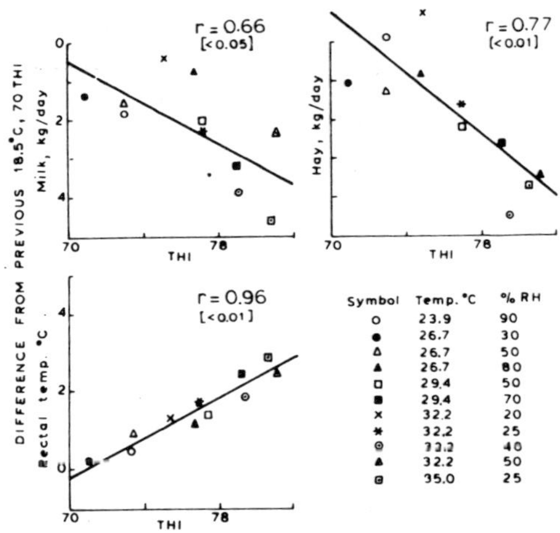

Figure 5. Seasonal heat effects in a temperate climate (Missouri) on conception rates.

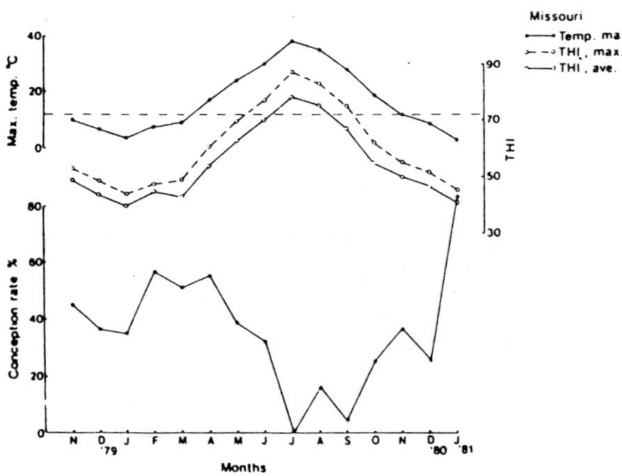

Table 2. Effect of tropical climate complexes on milk production of Holstein and Jersey cows.

Region	Annual Milk Production kg	Lactation Length days	Daily Milk Production kg	THI Animal (D)
Holstein				
United States[1]	7715	"305"	25	--
Arizona[1]	8331	"305"	28	66
Missouri[1]	6972	"305"	23	54
Puerto Rico[1]	4485	"305"	14.7	75
Mexico, Veracruz[2]	3534	"325"	10	73
Mexico, Tabasco[2]	2745	"305"	9	74
Egypt	----	----	9	69
Guyana	----	----	6	77
Jersey				
Missouri	----	----	16	54
Mexico, Veracruz[2]	2537	"318"	7.9	72
Costa Rica, CATIE[3]	2218	"300"	7.7	71

[1]Wiggins, G.R., C.A. Ernst, U.S.D.A. Summary, 1987, Herd Averages, Beltsville, Maryland. Official National Cooperative Dairy Herd Improvement Program (NCDHIP), 1987.

[2]Roman-Ponce, H. 1987. Lactation of Dairy Cattle in Humid Tropical Environments. Chapter 6, pp. 81-90. From: Bioclimatology and the Adaptation of Livestock. H.D. Johnson, Editor. Elsevier Publishers, Amsterdam.

[3]Costa Rica, CATIE, 1986. Rolling Herd Average for Jersey.

MILK YIELDS AND REPRODUCTIVE PERFORMANCE AS AFFECTED BY ENVIRONMENTAL HEAT AND THI

The effects of THI on milk yields of Holstein dairy cattle have previously been summarized by Johnson *et al*. (1961, 1962) and Johnson (1987b). The influence of environmental heat and THI is especially critical to conception rates of temperate zone lactating cattle during summer heat (Figure 5; Rabie, 1983) and in the subtropics (Ingraham *et* al., 1976). Most evidence suggests that reproductive failures associated with hyperthermia in cattle are due to embryonic death (Thatcher and Roman-Ponce, 1980; Putney et al., 1988) rather than insufficient LH, high prolactin or progesterone, which are responsible for ovulation and fertilization actions. Embryonic death may be due to thermal or uterine environmental changes (including hormonal or immunological changes; Spencer, 1988).

ALLEVIATION OR PARTIAL ALLEVIATION OF PRODUCTION LIMITATIONS

Nutritional

Decreased feed intake and a resultant decline in metabolizable energy (ME) intake is a major problem for the exotic (temperate) breeds of cattle imported into the tropics.

Nutritional Modification

The composition of the diet is believed to be important in alleviating heat stress. There are, however, no reliable scientific guidelines for feeding cows in hot climates. Milk yields did not change significantly in earlier studies where animals were forced to eat diets containing various ratios of forage/concentrate or isocaloric diets in which the ratio of fibre was varied (El-Khohja, 1979) or fat was added (Moody et al., 1971). We do know that cattle under heat stress will reach a hyperthermic state and will refuse forage but continue to eat concentrate.

Minerals

Since cows reduce their voluntary feed intake during hot temperate season weather and in the tropics (Collier et al., 1982), their mineral intake may also be less than optimal in hot weather, adding an additional limiting factor in hot humid environments. Kamal and Johnson (1978) also found negative mineral balances of cattle in which ration and total (urine and faeces) excretion were analyzed for Na, K, Ca, Mg, Zn, Cu, Fe, Co, Mo, and P.

Hormonal

Numerous hormones that are depressed in hot or tropical climates may warrant consideration as a means to prevent or restore milk yields of dairy cattle. Most promising is the bovine somatotropin (BST) which may soon be available commercially as a recombinant hormone. Prolonged heat stress was shown by our laboratory to lower plasma levels of growth hormone. Thus, assuming over-compensation may have occurred, the supplementation of growth hormone would increase milk yields and efficiency of energy conversion in hot climates. Recently Johnson et al. (1988), using the recombinant BST, increased milk yields under summer farm conditions by 18% and, in a subsequent laboratory simulated thermoneutral environment, by 25% and summer heat conditions by 26% over controls (Figure 6). The increases in milk, feed intake and metabolism did not increase body temperature more than controls due possibly to increased heat loss and/or efficiency of energy utilization (Johnson et al., 1987) (Figure 7).

Environmental (Shelter) Modification

Technology to avoid solar heat loads or increase heat losses from the animal to maintain heat balance is especially important for exotic temperate cows introduced into the humid tropics and during temperate zone summers. Shades have been shown by many scientists to minimize incoming radiation as much as 30% for the dairy cow and thus reduce heat loads (Roman-Ponce et al., 1977; Wiersma et al., 1984). Even in humid climates water sprays and high intensity fans can greatly improve milk yields (Igono et al., 1987, Johnson et al., 1987).

The effects of various temperature, humidity, wind combinations on milk yield and related heat balance measures were measured on lactating Holstein cows (Figure 8).

Genotype Modification: Production Adaptability Measures

Our goals are to provide the optimal micro-climate and micro-environment for the animal genotype, identify the genotype for adaptability as well as production potential and modify the genotype either by hormonal therapy (see above), or selective breeding. There are numerous examples of individual cows that are more heat tolerant and productive when subjected to heat stress (Johnson et al., 1962; Johnson, 1965; Johnson et al., 1967; Johnson, 1967; Johnson, 1987). Thus, selection can offer the potential to increase milk yield/cow in an existing microclimate, especially if selection includes production and adaptability indices (Horst, 1983; Johnson et al., 1988). These indices should improve stress resistance and avoid production compensation and excessive acclimatization.

A recent laboratory study on 51 cows at each stage of lactation demonstrated again a wide range of the ability of the individual cow with similar levels of milk yields (at thermoneutral) to produce when subjected to heat stress (Johnson et al., 1987) (Figure 9). These data clearly demonstrate that the milk yield and food intake of animals producing about 15-25 kg milk/day at thermoneutral (18°C) and were in the heat-tolerant portion of the population (Figure 9) declined less than the cows in the heat-sensitive portion of the distribution curve. This relationship of rectal temperature to performance (milk yield and feed intake) clearly describes the functional significance of thermal balance and energy-related functions.

Figure 6. Milk yields of control and BST-treated cows during summer farm and laboratory simulated TN and heat conditions.

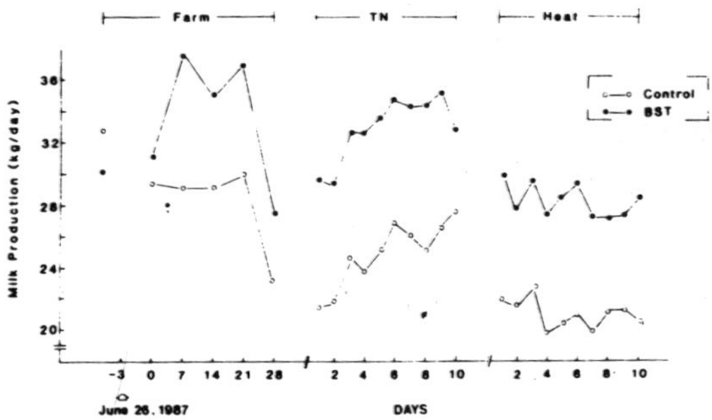

Figure 7. Milk production and rectal temperature of control and BST-treated cows.

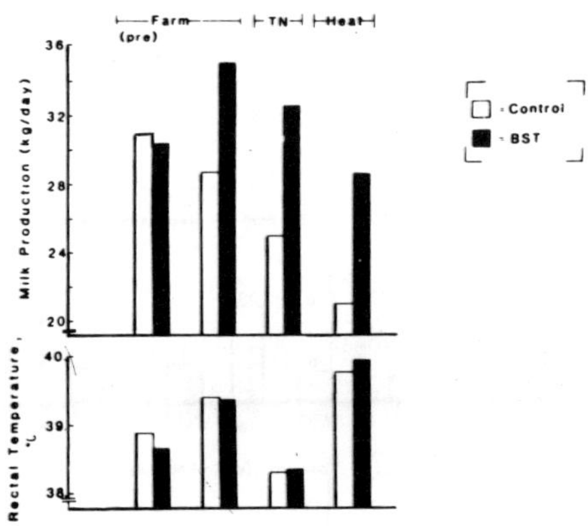

Figure 8. Changes in rectal temperature (°C) and % changes in milk yields under laboratory simulated and various combinations of temperature, humidity and wind. Control or base condition was 20°C, 40% RH and wind at 0.5 m/sec (MLS).

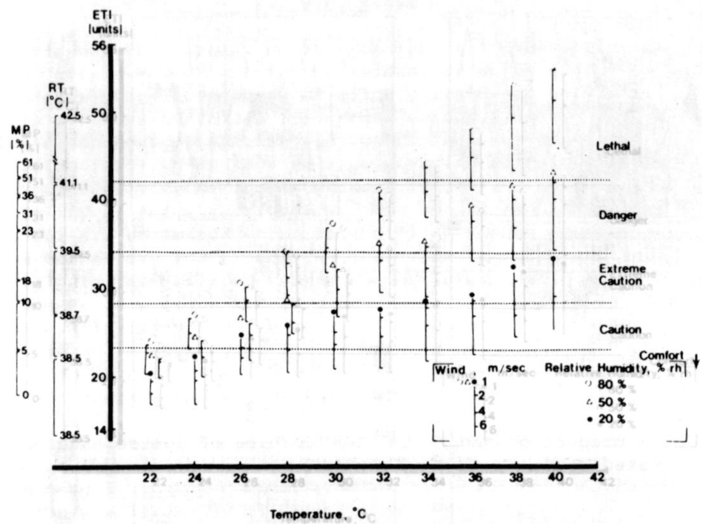

Figure 9. Frequency designations (-), (+) or (Intermediate) for productive adaptability indices (R = rectal temperature, °C; MP = milk production, % decline/day; and F = feed intake, Mcal/day).

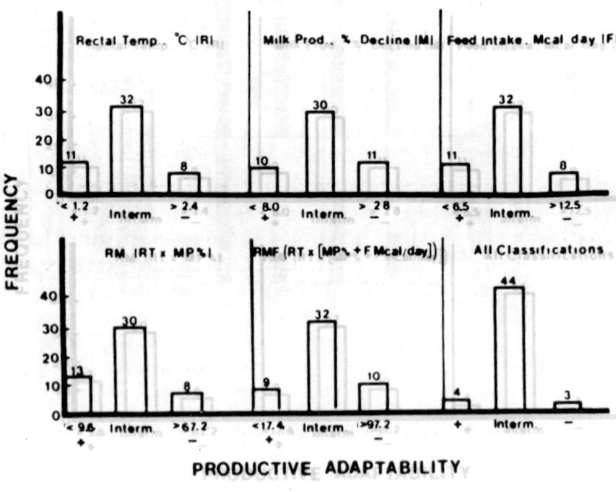

REFERENCES

Collier, R.J., Beede D.K., Thatcher W.W., Israel L.A. and Wilcox C.J. 1982 Influences of environment and its modifications on dairy animal health and production. J. Dairy Sci. 65:2213.

El-Kohja, M. Effect of environmental temperature on lactating dairy 1979 cows fed high and low fibre rations. M.S. Thesis, University of Missouri.

Horst, P. The concept of "productive adaptability" of domestic 1983 animals in tropical and subtropical regions. J. S. African Vet. Assoc. 3:159.

Igono, M.O., Johnson H.D., Steevens B.J., Krause G.F. and Shanklin 1987 M.D. Physiological, productive, and economic benefits of shade, spray, and fan system versus shade for Holstein cows during summer heat. J. Dairy Sci. 70:1069.

Ingraham, P.H., Stanley R.W. and Wagner W.D. Relationship of 1976 temperature and humidity to conception rate of Holstein cows in Hawaii. J. Dairy Sci. 59:2086.

Johnson, H.D., Kibler H.H., Ragsdale A.C., Berry I.L. and Shanklin 1961 M.D. Role of heat tolerance and production level in response of lactating Holsteins to various temperature-humidity conditions. J. Dairy Sci. 44:1191.

Johnson, H.D., Ragsdale A.C., Berry I.L. and Shanklin M.D. Effect of 1962 various temperature-humidity combinations on milk production of Holstein cattle. MO. Agric. exp. Sta. Res. Bulletin No. 791.

Johnson, H.D. Response of animals to heat. Meteor. Monographs. 1965 6:109.

Johnson, H.D., Hahn L., Kibler H.H., Shanklin M.D. and Edmonson H.D. 1967 Heat and acclimation influence on lactation of Holstein cattle. MO. Agric. Exp. Sta. Res. Bulletin 916.

Johnson, H.D. Environmental temperature effects (including humidity, 1976 radiation and wind) on animal functions and processes. IN: The Effect of Weather and Climate on Animals. Vol. 1, Part 1. S.W. Tromp (Editor-in-Chief) and H.D. Johnson (Editor). Swets and Zeitlinger BV, Amsterdam.

Johnson, H.D. Depressed chemical thermogenesis and hormonal function 1980a in heat. In: Environmental Physiology Aging, Heat and Attitude. Horvath and M.K. Yousef (Eds.), Elsevier Publishers, North Holland; New York.

Johnson, H.D. Environmental management of cattle to minimize the
1980b stress of climatic changes. Int. J. Biometeorol. 24:65-78.

Johnson, H.D. Bioclimates and Livestock. IN: Bioclimatology and the
1987a Adaptation of Livestock. Chapter 1. Elsevier Science Publishers, Amsterdam. pp. 3-16.

Johnson, H.D. Bioclimate Effects on Growth, Reproduction and Milk
1987b Production. In: Bioclimatology and the Adaptation of Livestock. Chapter 3. Elsevier Science Publishers, Amsterdam. pp. 35-57.

Johnson, H.D., Li R., Spencer K.J., Manalu W., Meador N.F. and Katti
1987 P.S. Influence of water sprays and wind in hot humid environments on milk yield, metabolic and thermo-regulatory functions. 18th Conf. Biometeor. and Aerobiology, Amer. Meteor. Soc., Sept.

Johnson, H.D., Shanklin M.D. and Hahn L. LXXV. Productive
1987 Adaptability of Holstein Cows to Environmental Heat. Part I. MO. Agric. Exp. Sta. Res. Bulletin 1060.

Johnson, H.D., Becker B.A., Spencer K.J., Collier R.J. and Baille C.A.
1988 Effect of field and laboratory heat stress on milk and physiological response of lactating dairy cows supplemented with Sometribove (Bovine Somatotropin, BST). J. Dairy Sci. Mtgs., Edmonton, Alberta.

Kamal, T.H. and Johnson H.D. Effect of high environmental temperature
1978 and age on trace elements metabolism in cattle. J. Middle Eastern Regional Radioisotope Centre.

Moody, E.G., Van Soest P.J., McDowell R.E. and Ford G.L. Effect of
1971 high temperature and dietary fat on milk fatty acids. J. Dairy Sci. 54:1457.

Putney, D.J., Drost D. and Thatcher W.W. Embryonic development in
1988 dairy cattle exposed to elevated ambient temperatures between days 1-7 post-insemination. Biol. Reprod. 34:100.

Rabie, H.M. Climatic effects on reproduction of dairy cattle. M.S.
1983 Thesis, University of Missouri.

Roman-Ponce, H., Thatcher W.W., Buffington D.E., Wilcox C.J. and Van
1977 Horn H.H. Physiological and production responses of dairy cattle to a shade structure in a sub-tropical environment. J. Dairy Sci. 60: 424.

Spencer, K.J. Influences of heat stress and related endocrine and
1988 uterine environmental changes on embryonic survival. M.S. Thesis, University of Missouri.

Thatcher W.W., and Roman-Ponce H. Effect of climate on bovine
1980 reproduction and lactation. J. Dairy Sci. 57:360.

Wiersma, F., Armstrong D.V., Welchert W.T. and Lough O.G. Housing
1984 systems for dairy production under warm weather conditions. World Anim. Rev. 50:16.

PHYSIOLOGICAL CONSTRAINTS TO MILK PRODUCTION: FACTORS WHICH DETERMINE NUTRIENT PARTITIONING, LACTATION PERSISTENCY AND MOBILIZATION OF BODY RESERVES.

by

Y. Chilliard

Bovine milk yield is related to both intrinsic genetic and extrinsic nutritional and environmental factors. Milk composition is related more to genetic factors but is also linked, in part, to extrinsic ones.

On a short-term basis, the efficiency of nutrient use for milk production is primarily dependent on the milk production level. As milk yield increases, a lower proportion of total feed intake is used for maintenance (a non-productive requirement that is more or less constant) of the cow. A cow producing 12 kg/d of milk is using about 50% of available nutrients for milk synthesis, whereas the corresponding value is 66% when milk yield increases to 22 kg/d.

This point needs however to be reconsidered in the context of developing countries, especially in the hot and humid tropics. In these countries, several factors limit the use of high-yielding dairy cows :

- highly digestible forages (and fertilizers used to produce them) are not available;
- cereal and other feeds of high nutritive value are not available in excess of what is needed for human or monogastric animal consumption, or are not available at an economic price;
- underfed specialised dairy cows decrease their milk production, but not enough to avoid excessive body weight loss, health and reproduction problems and even mortality;
- specialised high-yielding dairy breeds are not well adapted to climatic stress, to poor management and to endemic diseases and parasitism;
- zebu or crossbred dual-purpose cattle and buffaloes are well adapted to tropical conditions, produce in some cases 1000 - 3000 litres of high-fat milk per lactation and can be used as draught animals (see Preston and Leng, 1987 and Roman-Ponce, 1987, for complete analysis).

The present paper focuses on current knowledge of the physiological aspects of nutrient partitioning in lactating cows. Most data were obtained in high producing dairy cows from temperate countries. Therefore they do not apply directly to most milking cattle that are used in the tropics.

BODY GROWTH AND MAMMOGENESIS IN HEIFERS

Underfeeding of female dairy calves between birth and the first calving will decrease body growth, body reserves and will increase the rate of culling of the cows later (see also paper by J. Ugarte). Underfeeding during the first 4 months is more dangerous because there is little possibility of compensatory growth (see Johnson, 1988, and Troccon and Petit, 1989, for reviews). On the other hand, both underfeeding and overfeeding of heifers between birth and puberty can be detrimental to the milk production of the future cow (Figure 1).

Figure 1. Milk yield in 250 days of first lactation in relation to average daily gain before puberty (from 90 to 325 kg •) and after puberty (from 325 kg to first calving *). (Foldager and Sejrsen, quoted by Johnson, 1988)

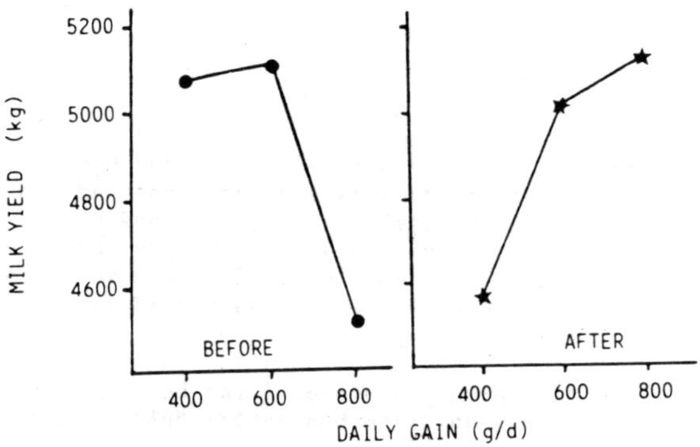

The negative effect of overfeeding has been related to an excessive development of mammary adipose tissue at the expense of parenchymal tissue growth, which is allometric during this period. Overfeeding is also known to decrease age at puberty, and thereby the length of the allometric phase of mammary development. These results have also been related to a decrease in somatotropin (BST or growth hormone) secretion during overfeeding and there are data showing that injections of BST can increase the growth of mammary parenchyma in heifers before puberty. There are, however, no data showing that BST treatment of heifers will increase the future milk yield of the cow.

From puberty to first calving, an increase in feeding level is generally favourable to subsequent milk yield as well as to the overall growth and deposition of body reserves in heifers. After calving, primiparous cows are simultaneously growing and producing milk. This could explain why body lipid deposition after lactation peak is lower in primiparous than in multiparous cows (Table 1).

Table 1. Body fat (kg) in Holstein and Friesian dairy cows[1]

		Peri-partum	Lactation peak	Mid lactation	Late lactation Dry period
Slaughter data[2]					
Robelin and Chilliard, 1983		122	41	-	35-122
Butler-Hogg et al., 1985		72	40	34-72	144
Dilution space data[3]					
Chilliard et al., 1984		81-90	51-68	90-96	-
Martin and Ehle, 1986		89-123	73	95	-
Vérité and Chilliard	m	104	72	-	104
(unpublished)	p	113	89	-	86

[1]See Chilliard (1987) for references.
[2]Dissectible adipose tissues
[3]Lipids estimated in vivo from body weight and deuterium water dilution space
m = multiparous cows
p = primiparous cows

PHYSIOLOGICAL ASPECTS OF MAMMARY SECRETION

Mammary secretion is a function of the number of milk secreting cells and the synthetic activity of each cell.

The last phase of mammogenesis (lobulo-alveolar growth) essentially takes place during the second half of pregnancy. It is genetically determined and controlled by oestrogens, progesterone, prolactin, BST and other hormones that are also implicated in the differentiation of the mammary cells into cells that are able to make milk (lactogenesis stage I). The onset of copious milk secretion (lactogenesis stage II) at parturition is due to elevated prolactin and adrenal steroids, simultaneous to progesterone withdrawal (Forsyth, 1983).

After calving, the metabolic activity of secretory cells is dependent on: (1) neuro-endocrine stimuli which are partly linked to suckling or milking. BST seems to be particularly important for the maintenance of lactation (galactopoiesis) in ruminants; (2) availability in arterial blood of nutrients that are used for milk synthesis (in most cases blood flow variations are consequences of tissue metabolic activities, but cardiac output could also be increased by prolactin and BST in non-lactating animals); (3) regular and complete evacuation of alveolar milk, to decrease intra-mammary pressure and to remove inhibitors that are secreted into the milk (feedback inhibition) (Mepham, 1983).

The decrease in milk yield after lactation peak (that determines milk persistency) results primarily from a decrease in the number of secreting cells. There is little knowledge on the possibility of manipulating the number of secretory cells during lactation. During extended lactation in the mouse, a stronger milking stimulus caused by new younger pups was able to increase the longevity of secretory cells, thus maintaining the number of cells at peak values and milk yield at two-thirds of peak values (suggesting that better milk persistency was due to cell number maintenance, whereas their metabolic activity decreased) (Knight et al., 1988).

Milk yield of dairy cows is clearly greater (25-40 %) when they are suckling their calves twice daily than when machine-milked twice daily. This could be due to a decrease of residual milk and/or to a better response of galactopoietic hormones to suckling (see Perez et al., 1985). Interestingly, when dairy cows suckle only during the first two months of lactation, they maintain an increased milk yield (above controls) after weaning, suggesting that the number of secreting cells was increased or that there was a carry-over effect on stimulating mechanisms (Everitt and Phillips, 1971). This can be relevant to simultaneous milking and suckling in dual-purpose herds.

Hemi-mastectomy during lactation is followed by compensatory yield that is partly due to an increase in cell numbers in the remaining gland. During thrice-daily milking in goats, milk secretion was increased in the short-term (hours or days) by removal of chemical feedback inhibitor and increased metabolic activity, and in the long-term (months) by increased cell number (resulting either from increased cell proliferation or from decreased cell death rate). The latter is however in contradiction with results in cows previously milked thrice-daily over 20 weeks, in which the increased yield was not maintained when they returned to twice-daily milking. The same observation can also be made after removal of long-term BST treatment

(Knight *et al*., 1988).

During concurrent pregnancy and lactation, there is a sharp decrease in milk yield during late pregnancy (after about 5 months in the cow), due primarily to increased oestrogen secretion that inhibits milk synthesis (and to some extent to competition for nutrients by the foetus). There is however at the same time a large proliferation of new secretory cells that will produce more milk during the following lactation (Knight *et al*., 1988). This proliferative phase is probably stimulated by drying-off the animals before the next lactation (Mepham, 1983). Hormonal induction of lactation in goats (without pregnancy) leads to lower milk yield and higher persistency, without change in mammary cell metabolic activities (Chilliard *et al*., 1986), probably due to lower mammogenesis and better cell maintenance.

Milk persistency between lactation peak and late pregnancy is also related to sustaining the metabolic activity of secretory cells (see above). It will be clearly related to adequacy of feeding (see below) and management (milking or suckling in good conditions) which enables expression of mammary cell secretory potential.

Under favourable conditions, unbred cows produce each month 94% of their yield during the preceding month. During a 5-year lactation, the yield of the 5th year was about 50% of that during the first (Smith, 1959). The same annual yield could be performed either with higher peak yield and lower persistency, or the other way round. Low persistency could be inherited or due to underfeeding or exhaustion of body reserves (see below) or to other unknown mechanisms that compensate for the higher solicitation of the mammary gland at peak yield. Persistency is indeed generally lower in higher producing animals, even if they are well fed (see Broster and Thomas, 1981 and Faverdin *et al*., 1987).

BODY RESERVES AND LACTATION IN THE DAIRY COW

Body fat at calving comprises 80-120 kg in "normally" fed Holstein x Friesian adult cows. The major part (but not all) of body lipids are stored in adipose tissues and can be lost during prolonged underfeeding. In well fed cows, the body fat that was lost after calving can be deposited again during declining lactation (Table 1).

Body proteins amount to 80-90 kg in Holstein cows, but most of them are structural components of the body and cannot be mobilized without irreversible degradation of the cow's potential. In underfed lactating cows, body protein loss estimated after slaughter or by deuteriated water did not exceed about 15 kg (20% of body protein), half of which was from muscles (Chilliard and Robelin, 1983). Body protein deposition in dry cows during fattening is lower than body protein loss in lactating underfed cows (Chilliard *et al*., 1987). This could explain why the muscle/bone ratio decreases in old cows

(Robelin, unpublished data). A great part of body water variations is linked to body protein variations. Liver glycogen reserves are very limited and used on a short-term (day-to-day) basis.

The extent and duration of body fat and protein mobilization after calving depends on milk potential, feeding level and quality, and initial body condition.

Effect of milk potential:

When milk potential increases (above 20 kg/d at peak yield), cows eat more feed but not enough to meet the mammary needs for nutrients, even with high quality diets fed *ad libitum*. Consequently, body weight loss (including lipids, water and protein) increases. Energy deficit lasted about 4 and 8 weeks in cows with peak milk yield of 20 and 40 kg/d respectively (Faverdin *et al*., 1987). However, when comparing different breeds of cows with the same milk yield, it was suggested that body weight loss was lower in dairy breeds (Holstein, Normandy) because of their higher feed intake capacity (Journet, Colleau and Piton, unpublished data). In respiratory chamber experiments with cows receiving 95% of their theoretical *ad libitum* intake, body energy loss was linearly related to milk production levels between 20 and 45 kg/d (Vermorel, Rémond and Vérité quoted by Chilliard *et al*., 1983).

In well-fed, high-producing dairy cows (more than 30 kg/d at peak yield), body fat loss was 30-60 kg during the first two months of lactation and body protein loss was 1-8 kg. In low-producing (7 kg/d), well-fed Hereford x Friesian cows, there was no significant body reserve mobilization (review by Chilliard *et al*., 1987). High producing dairy cows are able to maintain their milk yield only if their calculated protein deficit is lower than 10 kg during the first two months of lactation. This is in keeping with body composition data on body protein losses during *ad libitum* or restricted feeding (see above).

Effect of feeding level:

The immediate milk output response to metabolisable energy intake is curvilinear and follows the law of diminishing returns (Figure 2). When intake increases above mean theoretical requirements for cow milk potential, the milk yield response is lower and supplemental energy is stored as body reserves. In underfed cows milk yield does not decrease in proportion to energy intake, due to the use of mobilized body reserves for milk synthesis. Curves in Figure 2 show that for fixed (low) amounts of feeds, high-yielding cows (that are more underfed) are more responsive to supplementary feeding than low-yielding ones. It is also probable that at a very low feeding level there is little difference in the actual milk yield of cows with different milk potentials, although further investigations concerning this point are needed.

In Holstein x Friesian cows with the same milk potential (28-29 kg/d) a feed restriction (25% below <u>ad libitum</u> intake) decreased peak milk yield (about -2.5 kg/d) and increased body fat (-12 kg) and protein (-5 kg) losses during the first two months. The same restriction in cows with higher potential (33 kg/d) further decreased yield (-1.0 kg/d) and increased fat (-10 kg) and protein (-7 kg) losses, when compared with restricted cows of lower potential (Chilliard <u>et al</u>., 1983).

Hereford x Friesian cows (11-13 kg/d at peak yield) in good body condition that were fed at maintenance level for 6 weeks after 70 days of lactation produced 7-8 kg of milk per day, and body fat loss was estimated to be 22-23 kg. Body fat could be deposited again, and milk yield maintained, if cows were refed at 2 x maintenance requirements for 6 weeks (Topps <u>et al</u>., quoted by Chilliard, 1987).

Figure 2. The response of low and high yielding cows to an additional input of energy.

$r_1 > r_2$ (Broster and Thomas, 1981)
$r_3 \approx r_4$

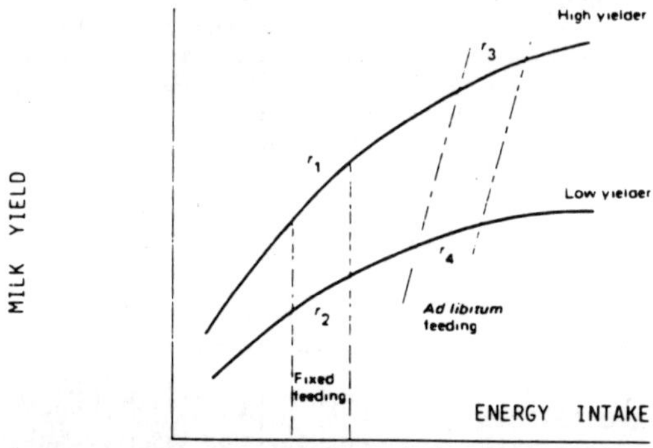

When cows are moderately underfed during early lactation, they are able to produce again as much milk as control cows after peak yield, if feed allowances are sufficiently high (Broster and Thomas, 1981 ; Coulon <u>et al</u>., 1987; Table 2), suggesting that mammary potential was not irreversibly altered. More generally, milk persistency is affected by the level of underfeeding after peak yield.

Table 2. Effect of body condition at calving and feeding level on dairy cow performance and body reserves (Rémond, Chilliard and Larnicol, unpublished).

Body condition at calving:	Fat		Lean	
Dietary energy after calving[1]	High	Low	High	Low
Condition score at calving	3.6	3.5	1.4	1.6
Dry matter intake (kg/d)	19.8	15.1	18.8	14.7
Milk yield (kg/d)	30.4	28.9	28.9	26.0
Milk fat content (g/kg)	40.8	43.6	39.7	39.1
Milk protein content (g/kg)	32.2	30.4	32.4	31.0
Energy balance (Mcal/d)	-0.55	-8.71	+1.21	-4.49
Plasma free fatty acids (mM)	0.48	0.94	0.47	0.63
Adipose cell diameter (μm)	-8	-21	-8	-15
Body weight change (kg)[2]	-32	-59	+11	-11
Condition score change	-0.2	-0.6	+0.7	0.0
Milk yield (kg/d over weeks 19-44)[3]	15.6	16.6	16.2	14.7

Data are from 51 cows (11 to 14 per group) between weeks 1 to 8 of lactation.

[1]Protein (PDI, protein digestible in the intestine) concentration was increased in "low" energy diet in order to achieve similar PDI intakes in "high" and "low" groups.

[2]Corrected for differences in dry matter intake.

[3]45 cows only (8 to 14 per group)

Effect of body condition at calving:

Body condition scoring allows the assessment of subcutaneous fat variations by manual palpation of the tail head and the loin areas. With a 0-5 scale, in 49 Holstein x Friesian slaughtered cows an increase of one unit of condition score was equivalent to 35 kg of body weight (r=0.69) and 28 kg of body lipids (r=0.85) (Rémond et al., 1988).

Body condition at calving is the result of body reserve mobilization and deposition cycles during the life of the cow, and more particularly during the previous lactation and dry periods. Increasing the level of feeding before calving generally increases subsequent milk yield (Broster and Thomas, 1981). This could be due to better mammogenesis and lactogenesis during the last weeks of pregnancy and the first days of lactation, to better digestive adaptation, as well as to better body condition (availability of body reserves) of the cow.

In well fed cows, body condition at calving has little effect on milk production, if the condition score is above 2. Fat cows generally have lower voluntary feed intake but produce the same amount of milk, due to body lipid (and probably protein) mobilization (review by Garnsworthy, 1988). These cows are however more susceptible to metabolic disorders (see following sections) and reproductive problems.

Differences between fat and lean cows are more pronounced during underfeeding. Underfed fat cows maintain their milk and fat yields due to a very high body lipid mobilization (Table 2). The decrease in milk protein yield and content is probably related in part to the low ability of body proteins to be mobilized. On the other hand, underfed lean cows still mobilize body reserves but not sufficiently to maintain their yield of milk, fat and proteins. In our trial they presented no higher incidence of health problems. When fed ad libitum from the 9th week, they reproduced normally and produced as much milk as fat cows from the "high" group during the second part of the lactation (Table 2). The excellent performances of lean cows that were well fed with high concentrate diet should also be underlined.

Body fat mobilization is also related to protein feeding. An increased supply of limiting amino-acids to the mammary gland can indeed increase milk yield and therefore the energy requirement of the cows. This can lead either to increased feed intake (with highly digestible forages) or to increased body fat mobilization that will supply fatty acids for fat synthesis and oxidative energy in the mammary gland (Journet and Rémond, 1981). These hypotheses are confirmed by the fact that milk response to an increased protein supply was higher in fat than in lean cows (Garnsworthy, 1988). Peak yield increase could be, however, followed by decreased persistency

(Broster and Thomas, 1981).

A knowledge of body stores is very important for understanding the long-term effects of underfeeding during several lactations (see Broster et al., 1984). In Wiktorsson's trials, underfeeding during the first year did not affect peak yield but increased body weight loss and decreased persistency. On the contrary, continued underfeeding during the second year decreased peak and total yield, without further body weight loss (due to body reserve exhaustion, cf. Table 2). When cows were refed during the third year, nutrients were directed primarily to body gain and milk yield was not fully returned to control level.

REGULATION OF NUTRIENT PARTITIONING

Both exogenous and endogenous nutrients are used by the mammary gland. They are more available during lactation because of increased intake and digestion, as well as increased endogenous nutrient mobilization, or decreased competition, by other tissues and organs. The liver plays a key role in glucose production. Adipose tissue (muscles) can release or take up fatty (amino) acids, glucose and acetate. Mineral metabolism in the bones and gut is also involved.

During lactation, mammary metabolism is stimulated by galactopoietic hormones, among them somatotropin (BST) plays a central role. BST is also involved in the coordination of extra-mammary metabolism in order to ensure the priority of the mammary gland for nutrients (teleophoresis, see Bauman and Currie, 1980). Knowledge on the mechanisms of BST action has increased very rapidly thanks to the treatment of cows with recombinant BST. Such a treatment rapidly increases milk yield but the feed intake response is delayed for 6-8 weeks. During this period body reserves are mobilized, but can be deposited again after several months of BST treatment in adequately fed cows (review by Chilliard, 1988).

The primary effect of BST is to stimulate the mammary gland, probably via stimulation of somatomedin production. BST also decreases glucose and amino acid oxidation, at the expense of adipose tissue long-chain fatty acids, and stimulates liver gluconeogensis. Part of these adaptations is due to BST counteracting insulin effects in various tissues. Lowered somatomedin secretion is partly responsible for "BST resistance" in underfed animals (Gluckman et al. 1987).

Insulin secretion and tissue responses to insulin decrease in early lactating animals whereas glucagon secretion is maintained or increased. This favours liver glucose production and adipose tissue mobilization (that is also favoured by higher beta-adrenergic sensitivity) and decreases glucose and amino-acid utilization in adipose tissues and muscles, but not in the mammary gland (see

Chilliard, 1987). Thyroid hormones are also lowered during early lactation, possibly decreasing basal energy expenditure and protein turn-over (Aceves *et al*., 1985). The respective effects of teleophoretic hormones such as BST, and of the mammary drain of nutrients in metabolic and endocrine adaptations to lactation are not completely understood.

High-producing dairy cows have been selected for their ability to give a high metabolic priority to the mammary gland. If teleophoretic mechanisms overdo homeostatic regulation, several metabolic disorders can occur (milk fever, ketosis, steatosis, infertility, etc.), even in cows receiving high quality diets in temperate countries. This can partly explain why specialised dairy breeds are more subject to health problems under tropical conditions in developing countries, contrary to local or crossbred cattle whose milk yield decreases more rapidly when they are underfed (Preston and Leng, 1987). Low milk potential could be considered as one facet of genetic and phenotypic adaptation to unfavourable conditions. Adaptation to heat stress is accompanied by changes in numerous lactogenic, galactopoietic or homeostatic hormones (Aceves *et al*., 1985; Johnson, 1987 and present meeting).

Efficiency of nutrient use for milk yield depends on the balance between glucogenic, lipogenic and aminogenic nutrients that are absorbed from the digestive tract and on absolute needs of the mammary gland for each particular nutrient (see Preston and Leng, 1987 and present meeting). It depends also on the effects of the absorbed nutrients on the endocrine state, particularly the insulin/BST ratio, and on the physiological needs of the cow to recover its "normal" body condition (see Chilliard, 1987). The efficiency of digestible energy utilization is decreased by heat stress, resulting in increased nutrient requirement at the absorptive level under tropical conditions (Roman-Ponce, 1987). As ambient temperature increases above 21°C, the cow lowers heat production by decreasing feed intake and milk yield, although some individuals are more resistant (Johnson, 1987).

Maintenance requirement represents a major portion of energy needs for cattle production and is consequently a major factor in overall energetic efficiency. Maintenance requirement is higher in high-producing dairy cattle breeds during weight maintenance periods without lactation. This is probably linked in part to the greater development of visceral organs with high protein turn-over (digestive tract, liver, heart, etc.) that are involved in handling larger amounts of feeds, nutrients and blood during lactation. This could also partly explain an increased maintenance requirement in the same animal during lactation, or according to previous higher feeding level (see Ferrell and Jenkins, 1985).

CONCLUSION

Knowledge of high-yielding dairy cow physiology and nutrition has rapidly increased during the last decades. Although not directly useful for milk production in tropical conditions, data on the effects of suckling, milking, underfeeding, body reserves, endocrine regulation and nutrient partitioning could probably be used as a basis for the planning and discussion of applied research, as well as for developing basic research, directly related to tropical conditions. ln term of priorities, a particular attention has to be payed to feeding and management of late pregnant-early lactating cows, since they are more responsive at this physiological stage.

REFERENCES

Aceves, C., Ruiz-J, A., Romero, C. and Valceroe-R, C. Homeorhesis 1985 during early lactation. Euthyroid sick-like syndrome in lactating cows. Acta Endocrinologica 110 : 505-509.

Bauman, D.E. and Currie, W.B. Partitioning of nutrients during 1980 pregnancy and lactation: a review of mechanisms involving homeostasis and homeorhesis. Journal of Dairy Science 63: 1514-1529.

Broster, W.H., Clements, A.J. and Broster, V.J. Multiple lactation 1984 feeding of the dairy cow. World Review of Animal Production 20: 61-69.

Broster, W.H. and Thomas, C. The influence of level and pattern of 1981 concentrate input on milk output. In: Haresign W. (Editor). Recent Advances in Animal Nutrition. Butterworths: London, pp. 49-69.

Chilllard, Y. Revue bibliographique: Variations quantitatives et 1987 métabolisme des lipides dans les tissus adipeux et le foie au cours du cycle gestation-lactation. II Chez la brebis et la vache. Reproduction Nutrition Développement 27: 327-398.

Chilliard, Y. Review. Long-term effects of recombinant bovine 1988 somatotropin (rBST) on dairy cow performances. Annales de Zootechnie 37: 159-180.

Chilliard, Y., Delouis, C., Smith, M.C., Sauvant, D. and Morand-Fehr, 1986 P. Mammary metabolism in the goat during normal or hormonally-induced lactation. Reproduction Nutrition Développement 26: 607-615.

Chilliard, Y., Rémond, B., Agabrlel, J., Robelin, J. and Vérité, R.
1987 Variations du contenu digestif et des réserves corporelles au cours du cycle gestation-lactation. Bulletin Technique C.R.Z.V. de Theix, I.N.R.A. 70: 117-131.

Chilliard, Y., Rémond, B., Sauvant, D. and Vermorel, M. Particularités
1983 du métabolisme energétique. In "Particularités nutritionnelles des vaches à haut potentiel de production". Bulletin Technique C.R.Z.V. de Theix, I.N.R.A. 53: 37-64.

Chilliard, Y. and Robelin, J. Mobilization of body proteins by early
1983 lactating cows measured by slaughter and D_2O dilution techniques. IVth International Symposium Protein metabolism and nutrition (Clermont-Ferrand), European Association for Animal Production. Publication no. 31, Volume II, 195-198.

Coulon, J.B., Petit, M., D'Hour, P. and Garel, J.P. The effect of
1987 level and distribution of concentrate supplementation on performance of dairy cows. Livestock Production Science 17: 117-133.

Everitt, G.C. and Phillips, D.S.M. Calf rearing by multiple suckling
1971 and the effects on lactation performance of the cow. Proceeding New-Zealand Society Animal Production 31: 22-40.

Faverdin P., Hoden A., Coulon J.B. Recommandations alimentaires pour
1987 les vaches laitiéres. Bulletin Technique C.R.Z.V. de Theix, I.N.R.A., 70: 133-152.

Ferrell, C.L. and Jenkins, T.G. Cow type and the nutritional 1985 environment: nutritional aspects. Journal of Animal Science 61: 725-741.

Forsyth, I.A. The endocrinology of lactation. In Biochemistry of
1983 lactation, 309-349 (Mepham, T.B., Ed.) Elsevier Science Publishers B.V.

Garnsworthy, P.C. The effect of energy reserves at calving on
1988 performance of dairy cows. In "Nutrition and lactation in the dairy cow" (edited by P.C. Garnsworthy) pp. 157-170. Proceedings of the 46th University of Nottingham Easter School in Agricultural Science.

Gluckman, P.D., Breier, B.H. and Davis, S.R. Physiology of the
1987 somatotropic axis with particular reference to the ruminant. Journal of Dairy Science 70: 442-466.

Johnson, H.D. Bioclimates and Livestock. In "World Animal Science,
1987 B5. Bioclimatology and the adaptation of livestock" pp.3-16. Edited by Johnson H.O., Elsevier, Amsterdam.

Johnson, I.D. The effect of prepubertal nutrition on lactation
1988 performance by dairy cows. ln "Nutrition and lactation in the dairy cow" (edited by P.C. Garnsworthy) pp. 171-192. Proceedings of the 46th University of Nottingham Easter School in Agricultural Science.

Journet, M. and Rémond, B. Response of dairy cows to protein level in
1981 early lactation. Livestock Production Science 8: 21-35.

Knight, C.H., Wilde, C.J. and Peaker, M. Manipulation of milk
1988 secretion. In "Nutrition and lactation in the dairy cow" (edited by P.C. Garnsworthy) pp. 3-14. Proceedings of the 46th University of Nottingham Easter School in Agricultural Science.

Mepham, T.B. Physiological aspects of lactation. ln "Biochemistry of
1983 Lactation" (edited by Mepham T.B.) pp. 3-28. Elsevier Science Publishers B.V.

Perez, O., Jimenez de Perez, N., Poindron, P., Le Neindre, P. and
1985 Ravault, J.P. Relations mére-jeune et réponse prolactinique a la stimulation mammaire chez la vache: lnfluences de la traite et de l'allaitement libre ou entravé. Reproduction Nutrition Développement 25: 605-618.

Preston, T.R. and Leng, R.A. Matching livestock production systems to
1987 available resources in the tropics and sub-tropics. Penambul Books, Armidale.

Rémond, B., Robelin, J., Chilliard, Y. Estimation de la teneur en
1988. lipides des vaches laitiéres Pie Noires par la méthode de notation de l'état d'engraissement. INRA Production Animale 1: 111-114.

Roman-Ponce, H. Lactation of dairy cattle in humid tropical
1987 environments. ln "World Animal Science, B5, Bioclimatology and the adaptation of livestock", pp. 81-92. Edited by Johnson, H.D. Elsevier. Amsterdam, Netherlands.

Smith, V.R. Physiology of lactation. Fifth Edition. Iowa State
1959 University Press, Ames, Iowa, USA.

Troccon, J.L. and Petit, M. Croissance des génisses de renouvellement
1989 et performances ultérieures. INRA Production Animale 2: 55-64.

INFLUENCE OF NUTRITION ON REPRODUCTIVE PERFORMANCE OF THE MILKING/GESTATING COW IN THE TROPICS

by

K-H. Lotthammer

Good milk production and numbers of calves per unit time are only obtained by achieving early conception in heifers and a short inter-calving interval in adult cows. Factors affecting fertility in dairy cows are numerous (Table 1).

Table 1. Possible factors affecting fertility in dairy cattle.

A. GENETICS 20 per cent
B. ENVIRONMENT 80 per cent
- 1. CLIMATE - temperature, humidity
- 2. INFECTIONS
- 3. PARASITES
- 4. MANAGEMENT = FARMER
  - a. hygiene
  - b. HEAT DETECTION
  - c. suckling effect
  - d. FEEDING - NUTRITION (deficiency - excess)
    - Rearing calves - HEIFERS - adult cows
    - Nutrients - contents of feed plants and concentrates
    - ENERGY - Protein
    - minerals, trace elements
    - vitamins
    - substances in food with negative influences on fertility (nitrate, goitrogens, oestrogenics-antioestrogenics, mycotoxins)
  - e. Food production
    - Fertilisation, quality of soil
    - Cultivation and preparing of feed plants and pasture

C. MALE FERTILITY

It is estimated that about 80 per cent of the variance in fertility is due to environmental factors, of which more than 50 per cent is explained by nutrition, when severe infections and male fertility are excluded. Even predisposition to infectious diseases can be caused or increased by nutritional failures. Therefore balanced feeding is fundamental to milk production as well as health and fertility.

The present paper is mostly based on data which were obtained in dairy cows kept under temperate climate conditions, because of a lack of knowledge about relationships under tropical conditions. Except for some cases, the results should be seen as an indication of possible relationships.

In feeding dairy cows, three basic points must be considered in order to get good reproductive performance:

1. Balanced feeding is necessary throughout the year (lactation, gestation, dry period).
2. Reproduction can be affected by both an excess as well as a nutritional deficiency.
3. Interactions exist between the factors affecting fertility so that the combined effects are additive.

The last point mainly applies to energy deficiency. <u>Energy</u> is an important nutrient for dairy cows both before and after calving and there is no substitute for energy in the diet of ruminants. A balance of energy and protein is required, even before calving and in the dry period (Table 2).

Table 2. Reproductive performance and frequency of metabolic diseases in dairy cows fed different levels of energy and protein in the dry period.

	Group 1	Group 2	signif.
Feeding level:	Maintenance + 16 kg FCM <u>HIGH</u> %	Maintenance + 2 kg FCM <u>LOW</u> %	
delayed uterine involution	53.6	17.2	**
puerperal endometritis	70.8	26.9	**
follicular cysts	44.8	18.7	*
conception rate with one or two inseminations	46.4	74.1	*
paresis pueperalis	26.2	6.3	*
subclinical acetonaemia	65.5	45.5	*

Lotthammer and Faries (1975)

As table 2 shows, all parameters of reproductive performances in Group 1 (overfed during the dry period) pointed to lower fertility compared to Group 2 with restrictive feeding (for maintenance and 2 kg milk). Furthermore the incidences of sub-clinical ketosis and parturient paresis were higher. Milk production after calving however was not increased (difference 0.5 kg/day), but the milk fat content was significantly elevated by 0.9% in the overfed group due to lipolysis. This effect, moreover, results in fatty liver. The metabolic stress is caused by a decreased intake after calving. Similar results were also obtained by Boisclair et al. (1988) and Flipot et al. (1988). There is a consensus among the authors that overfeeding should be avoided in the last stages of lactation to prevent fattening. About two weeks before the expected calving date, the cow's rumen should be prepared by a gradual increase in concentrates (0.5 kg per day more).

On the other hand, an energy deficiency before calving (below maintenance) should be avoided as well because this leads already at this stage to metabolic stress with subclinical ketosis and liver damage, followed by a higher incidence of retained placenta, endometritis and low conception rates in the following lactation.

The negative effect of an insufficient energy provision before calving will be enhanced by an energy deficiency after the following parturition. Probably due to the liver damage which occurs in this case, there is also an impact on fertility (Lotthammer, 1975; Reid et al., 1979). One also finds more severe infections of the udder, mainly from E.Coli. Furthermore we found that, in herds with a high incidence of liver damage, the response to vaccination against IBR/IPV (BHV-1) was significantly depressed. These results suggest an immunosuppression due to liver damage.

The importance of energy after parturition is well known. Already in the first two to three weeks of lactation, energy from any source is important for the onset of ovarian activity (Butler et al., 1981; Terqui et al,. 1982; Villa-Godoy et al., 1988) and, related to this, for uterine involution. Energy deficiency leads to acyclia, silent heat, delayed ovulations and follicular cysts. Significant correlations exist between fertility and weight loss or body condition, as indicators of negative energy balance in the first weeks after calving (Oxenreider and Wagner, 1971; Godfrey et al., 1982; Rutter and Randel, 1984). Weigelt et al. (1988) also report embryonic mortality in cows with energy deficiency. The effects of energy intake on reproduction follow the pathway summarized in Figure 1. That even the result of therapeutic treatments of reproductive disturbances such as endometritis depends on the nutritional status is demonstrated in Table 3.

Under tropical conditions in the dry period, energy provision is insufficient and related to fertility problems. This is suggested by the field investigations of Betancourt et al. (1985) carried out in Columbia with the dual purpose Zebu (Figure 2). More exactly, we could demonstrate the correlations between energy and reproduction in relation to the seasonal conditions in Pakistan in milk buffaloes (Figure 3).

Figure 1. Effects of energy imbalance and acidosis ante- and post-partum on health and reproductive performance in dairy cows (Lotthammer 1987).

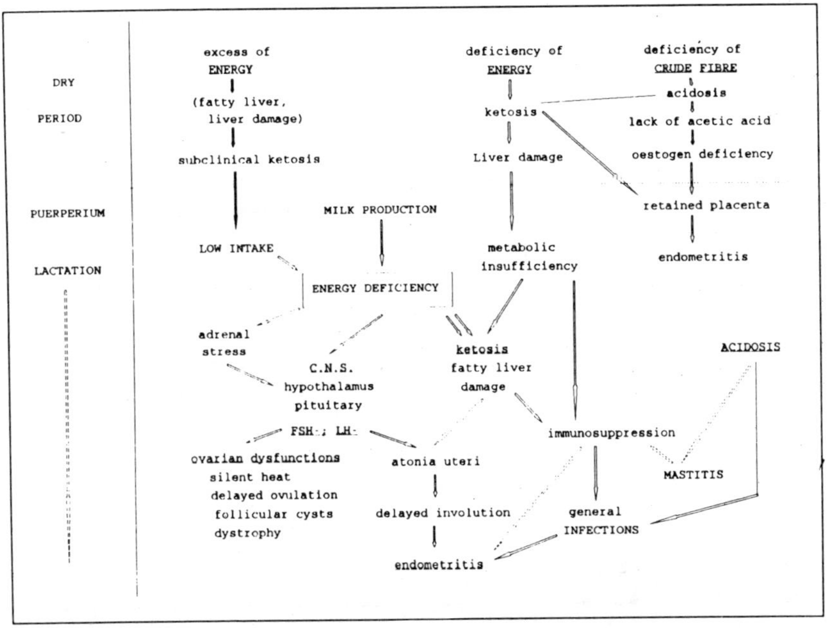

Figure 2. Monthly amount of rainfall in Cordoba (Colombia) and percentages of static ovaries in dual purpose Zebus in the dry (Jan to May) and rainy season (June to Dec) (Betancourt _et al_., 1985).

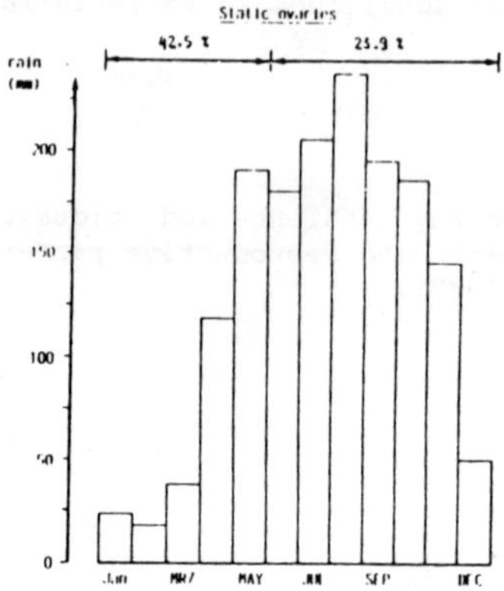

Table 3. Results in pregnancy, number of treatments and inseminations per pregnancy after treatment of puerperal and postpuerperal endometritis and losses in cows with different nutritional status and health.

Parameter	nutritional status/health normal	energy deficiency excess of protein	energy deficiency liver damage
% of pregnancies	94.4	81.0	54.5
treatments per cow	1.22	1.46	1.66
inseminations/pregnancy	1.39	2.19	2.59
interval treatment - conception (days)	48.50	66.90	68.30
costs of treatment/ conception ($)	17.-	24.-	40.-
differences to "normal" group by costs of treatment and losses per cow ($)		+ 78.-	+ 100.-

(Escherich and Lotthamer, 1987)

Figure 3. Monthly distribution of conceptions and concentrations of glucose and urea in blood serum of milk buffaloes in the course of the year (Bode, 1989).

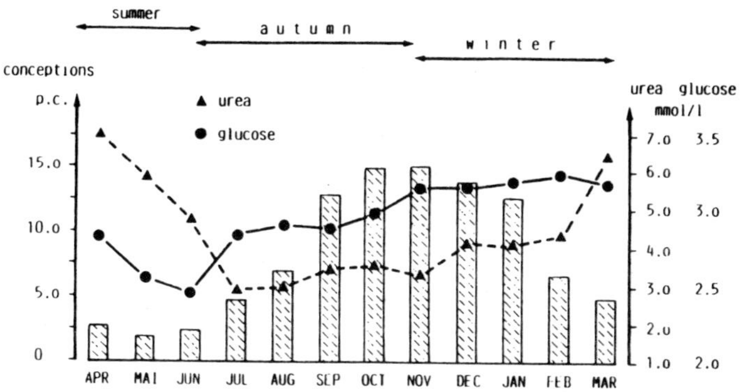

As Figure 3 demonstrates, the glucose levels in the (dry) summer are at a minimum and they increase with the beginning of rainfall in the autumn. At the same time, the percentage of conceptions increases too. This suggests that, with better feeding conditions following rainfall, the energy supply is also improved which results in a higher conception rate. The high urea levels in the dry period indicate imbalanced energy/protein supply too, due to an energy deficiency.

Figure 4 illustrates the problem of providing energy and protein in regions with a dry and rainy season. The nutritional conditions are changing between high intake and deficiency with consequences for metabolism. This mainly influences reproduction in the dry season. To get better results, efforts should be made to balance these conditions. That means transferring part of the surplus from the rainy season to the dry season by the following activities:

1. Ensilage of well grown feed plants (king grass, corn, etc.)
2. Introduction of new feed plants (corn or other) (Lotthammer, 1982)
3. Irrigation of grassland to shorten the dry period and to extend the period of vegetation
4. Good grassland management.

Figure 4. Schematic situation of energy and protein supply in regions with a rainy and dry season.

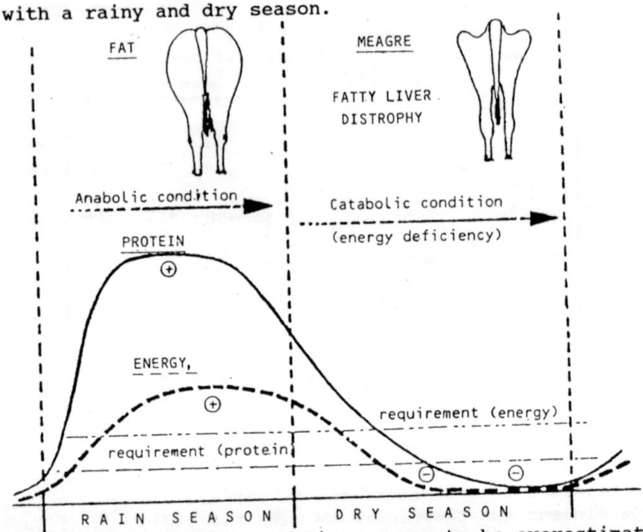

The importance of protein sometimes seems to be overestimated, if it is appreciated that ruminants are producing about 70% of their "own protein" by means of microbes in rumen. These need mainly energy in the form of carbohydrate. This means that, with normal rumen conditions for the microbes, ruminants are already supplied with protein and energy via gluconeogenesis. This could be demonstrated by Kaufmann (1976) and is indicated by the positive correlation between energy intake and protein content in milk.

We found a curvilinear correlation between conception rate and protein supply. Protein deficiency as well as an excessive protein supply causes acyclia and low conception rates.

The negative effect of a relatively excessive intake of protein is enhanced by energy deficiency at the same time. Under these conditions, a large amount of ammonia is flowing through the liver and causes liver damage. In these cases we find static ovaries, silent heat, anaphrodisia, purulent endometritis and embryonic mortality. This can be caused by alterations of uterine secretions (Jordan *et al*., 1981, Ferguson and Chalupa, 1989). Uterine treatments with antibiotics or other preparations in these cases are not very successful until nutrition is balanced (see Table 4) and therefore they are not economic.

Table 4. Incidence of diseases in health and fertility *post-partum* after grazing on pastures with different concentrations of nitrate in grass dry matter.

Disease	NO_{3^-} concentration in dry matter (%) < 0.30	0.30 - 0.50	> 0.50
Paresis	0	6.1	17.2
Puerperal endometritis	25.0	36.4	44.8
Retained placenta	0	13.6	10.3
Abortions/stillbirths	7.1	6.1	3.4

Minerals

For many years, phosphorus deficiency was given as a main cause of infertility. Legel (1970) demonstrated in a definitive experiment that phosphorus deficiency decreased total intake which caused a lower energy supply and lower weight gain in heifers. A negative effect on reproduction was not found. In the case of stress, the negative effect of P deficiency is enhanced. The Ca:P ratio also seems to be important. The results of our experiments show that the frequency of non-infectious endometritis is increased when Ca:P ratios decrease. Furthermore a lower content of manganese is found in uterine tissue.

This also demonstrates the importance of calcium, which has functions for the uterine performance, especially after calving for involution of the uterus. The advice is to keep Ca:P ratio in total intake over 2:1 with marginal P supply which should be higher under stress conditions. Depending on soil quality and fertilizer use, a deficiency of sodium and a correlated excess of potassium can reduce fertility by irregular oestrus cycles, endometritis and follicular cysts (Lotthammer and Ahlswede, 1973). The Na:K ratio should be kept under 10:1. Sodium supplementation by salt is very cheap and should be given *ad libitum*.

Of the trace minerals, manganese and selenium may influence reproduction. Manganese supply is correlated with pH value of the soil because high pH values inhibit uptake from the soil. A deficiency of manganese produces anaphrodisia, endometritis and abortions (Anke *et al*. 1987). Daily supply per cow should be 1000 mg. Recently selenium deficiency has been discussed, in combination with vitamin E, producing retained placenta, endometritis and cystic ovaries (Harrison *et al*., 1986). The authors state that only a combined treatment with selenium and vitamin E improves fertility.

Vitamins

Of the vitamins, vitamin A is seen nearly exclusively as the factor affecting fertility. In several experiments with heifers and milking cows, we could demonstrate a negative effect of ß-carotene deficiency on fertility, unrelated to vitamin A (Lotthammer, 1979). The failures are silent heat, delayed ovulations, follicular and luteal cysts, early embryonic mortality and diarrhoea in calves. These problems occurred in spite of sufficient vitamin A. The daily supply needed is 125 mg for dry cows and heifers and about 300 mg for milking cows. The carotene status is easily observed by the colour of the serum.

The effect of vitamin E deficiency cannot be separated from selenium deficiency (see above). Both must be considered to prevent fertility problems. The requirement will be supplied by green feed or silage.

Other factors

Under certain conditions, some substances in plants can affect health and fertility. As pointed out above (Table 4), fertility problems increased with higher concentrations of nitrate in grass. The content of nitrate in plants is positively correlated with dry conditions and fertilization with nitrogen. Energy deficiency enhances the negative effect. Also some plants (e.g. Crucifeiae) have high concentrations of nitrate.

Other substances in plants are oestrogenic. These substances can be produced by plants themselves (Trifolium) or by fungi on plants and fungi cause problems in tropical areas. The oestrogenic substances influence fertility directly via the ovaries in a very severe way (Kallela, 1968; Lotthammer *et al*., 1970). Avoiding the growth of fungi is the best way to prevent disturbances.

Conclusion

More research is needed to increase our knowledge of the effects of nutritional factors affecting fertility and health. Furthermore, more work is needed to determine the contents of all nutrients in local feeds, in relation to season and soil conditions.

REFERENCES

Anke, M., Groppel B., Reissig W., Lüdke H., Grün M. and Dittrich G.
1987 Manganese deficiency in ruminants: (3) Manganese stimulated reproductive, skeletal and nervous disturbances in female ruminants and their offspring. Archiv Tierernährung 23: 197-204.

Betancourt, A., Lobo C.A., Otte E. and Villafane F. Resultados de una
1985 encuesta sobre salud y produccion animal en Cordoba, Monteria (Colombia) 1982/1983/1984. Polycopy ICA/GTZ.

Bode, E. Ernährungs- und managementbedingte Einflüsse auf das
1989 Fruchbarkeitsgenchehen bei Nili-Ravi Büffeln im Punjab/Pakistan. Dissertation, Vet.med.Faculty, Freie Universität Berlin (West) (in preparation).

Boisclair, Y., Grieve D.G., Allen O.B. and Curtis R.A. The effect of
1988 prepartum energy, body condition, and sodium bicarbonate on health and blood metabolites of Holstein cows in early lactation. J. Dairy Sci. 70: 2280-2290.

Butler, W.R., Everett R.W. and Coppock C.E. The relationships
1981 between energy balance, milk production and ovulation in postpartum Holstein cows. J. Anim. Sci. 53, 742-748.

Escherich, J. and Lotthammer K.-H. Individuell und managementbedingte
1987 Einflüsse auf den Erflog con Sterilitätsbehandlungen bei Rindern. Zuchthygiene 22: 134.

Ferguson, J.D. and Chalupa W.. Symposium: Interactions of Nutrition
1989 and Reproduction. Impact of Protein Nutrition on Reproduction in Dairy Cows. J. Dairy Sci. 72: 746-766.

Flipot, P.M., Roy G.L., Dufour J.J. Effect of peripartum energy
1988 concentration on production performance of Holstein cows. J. Dairy Sci. 71: 1480-1850.

Godfrey, R.W., Rouquette F.M., Jr. and Randel R.D. The effect of cow
1982 weight change on cow reproductive performance. J. Anim. Sci. 55: 21.

Harrison, J.H., Hancock D.D., Pierre N.S., Conrad H.R. and Harvey W.R.
1986 The effect of prepartum selenium treatment on uterine involution in the dairy cow. J. Dairy Sci. 69: 1421-1425.

Jordan, E.R., Chapman T.E., Holtan D.W. and Swanson L.V. Alterations
1981 in mineral composition of uterine secretions and plasma from dairy cows fed high and low crude protein rations. Biol. Reprod. 24, (Suppl. 1): 146 A.

Kallela, K. The oestrogenic effect of red clover pasture on the
1968 ovariectomized heifer. Nord. Veterinærmed. 20: 185-182.

Kaufman W. Zur Bedeutung der Energieversorung hockleistender Milchkühe
1976 für den Milcheiweißgehalt und die Fruchtbarkeit. Keiler Milchwirtch. Forschungsberichte 28: 347-357.

Legel, S. Kritische Betrachtungen sowie Untersuchungen über den
1970 Einfluß von Phosphor auf das Fruchtbarkeitsgeschehen weiblicher Jungrinder. Fortpflanz., Besam, Aufzucht d. Haustiere 6: 244-255.

Lotthammer, K.-H. Eirstocks- und Gerbärmuttererkrankungen bei
1975 subklinschen Stoffwechselsörungen der Milchkühe. Prakt. Tierarzt 56: 24-29. (Collegium veterinarium).

Lotthammer, K.-H. Importance of ß-carotene for fertility of dairy
1979 cattle. Feedstuffs 51: 16-50.

Lotthammer, K.-H. Effect of improving energy intake during early
1982 lactation by corn-silage on metabolism and fertility of dairy cows. Current Topics in Veterinary Medicine and Animal Science 20: 409-412.

Lotthammer, K.-H. Fütterungsbedingte Einflußfaktoren auf die
1987 Fruchbarkeit des Rindes. AGRAR-Service Information 1987 No. 4: 8-14. Graz, Austria.

Lotthammer, K.-H., Grunert E. and Elghamry M.I.
1970 Fruchtbarkeitsstörungen und Fruchtbarkeit in Rinderbeständen. Dtsch. tierärztl. Wochenschr. 77: 445-448.

Lotthammer, K.-H. and Ahlswede L. Parotisspeicheluntersuchungen zur
1973 Beurteilung der Na- und K-Versorgung bei Rindern und ihre Beziehung zur Fruchtbarkeit. Tierärztl. Umsch. 28: 419-426.

Lotthammer, K.-H. and Faries E. The influence of nutrition (energy
1975 and digestible crude protein) on some blood parameters, health and fertility in late pregnant milking cows. Proc. 26th Annual Meeting of European Association of Animal Production in Warshaw, Poland.

Oxenreider, S.L. and Wagner W.C. Effect of lactation and energy intake
1971 on postpartum ovarian activity in the cow. J. Anim Sci. 33: 1026-1031.

Reid,I.M., Roberts C.J. and Manston R. Fatty liver and infertility in
1979 high-yielding dairy cows. Vet. Record 104: 75-76.

Rutter, L.M. and Randel R.D. Postpartum endocrine function and ovarian
1984 activity in cows. Current topics in veterinary medicine and animal science 20: 38-408.

Terqui, M., Chupin D., Gauthier D., Perez N., Pelot J. and Mauleon P.
1982 Influence of management and nutrition on postpartum endocrine function and ovarian activity in cows. Current topics in veterinary medicine and animal science 20: 384-408.

Villa-Godoy, A., Hughes T.L., Emery R.S., Chapin L.T. and Fogwell R.L.
1988 Association between energy balance and luteal function in lactating dairy cows. J. Dairy Sci. 71: 1063-1072.

Weigelt, B., Weigelt R., Barth T., Bach S., Eulenberger K. and Schulz
1988 K. Untersuchungen zur embryonalen Mortalität in einer Milchkuhherde. Monatshefte Veterinärmedizin 43: 157 -160.

THE ROLE AND MECHANISMS OF GENETIC IMPROVEMENT IN PRODUCTION SYSTEMS CONSTRAINED BY NUTRITIONAL AND ENVIRONMENTAL FACTORS

By

Ola Syrstad

INTRODUCTION

One of the many constraints on milk production in the tropics is the poor genetic potential of the indigenous animals. Tropical cattle are mostly of zebu (Bos indicus) type. These cattle are well adapted to the conditions prevailing in the tropics. Natural selection over hundreds of generations has provided them with a high degree of heat tolerance, some resistance to many tropical diseases and the ability to survive long periods of feed and water shortage. However, their dairy potential is poor; they have low milk yield, they are late maturing and usually do not let down milk unless stimulated by the sucking of the calf.

Genetic improvement alone might not result in drastic increases of milk production in the tropics, but it is a prerequisite for such increases. Genetically more productive animals are also the best incentive to improved feeding and management.

GENOTYPE AND ENVIRONMENT

The performance of an animal is the result of the joint action of its genotype and the non-genetic effects to which it is exposed. The non-genetic factors are often collectively termed the "environment".

The genotype is often conceived as a frame which restricts the performance to a given level. Below this level, the performance is determined by the environment. This concept is visualized in Figure 1a. Two genotypes, A and B, are considered. The superiority of the better genotype, A, is expressed only if the environment is more favourable than that which is necessary to exploit fully the potential of the poorer genotype, B. When the environment is worse than this, both genotypes would perform similarly and genetic improvement beyond the level of B would be of no use. According to this model both genotype and environment can act as bottlenecks which restrict performance. Although this might seem reasonable, the available evidence indicates that the concept is in general not providing an appropriate description of the interaction between genotype and environment.

The model illustrated in Figure 1b is much more likely to be correct in most cases. Here the superiority of the better genotype, A, is realized, regardless of the environmental conditions, but the

Figure 1. Different models of genotype x environment interaction. For explanation, see text.

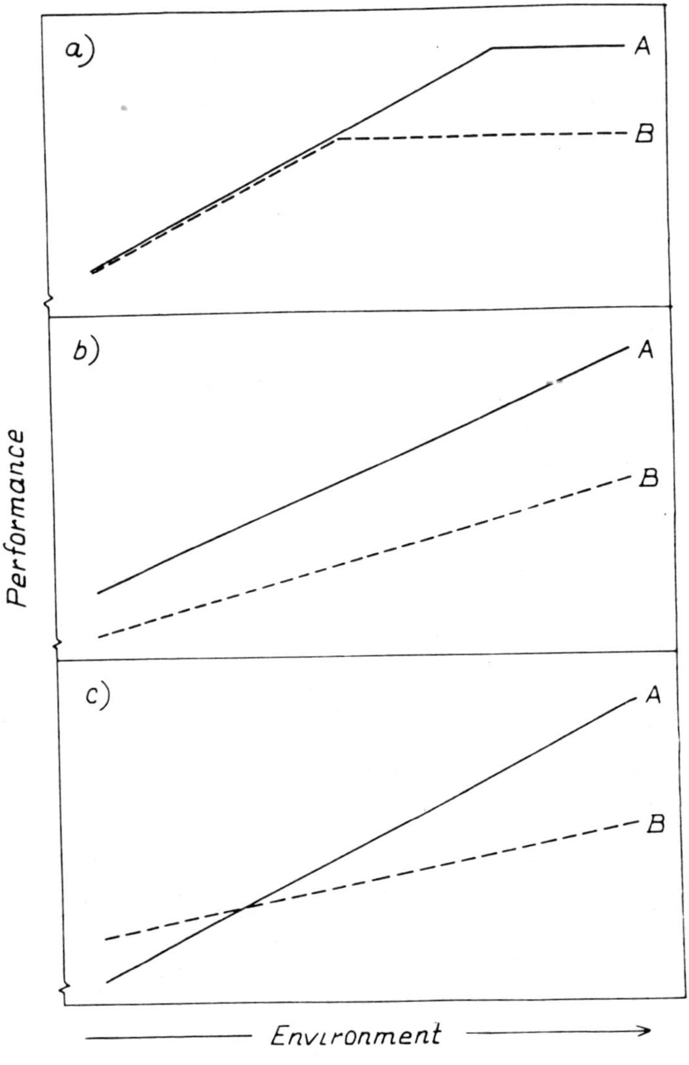

difference between the two genotypes increases as the environment improves. This means that genotype A responds more to improved conditions than genotype B (indicated by the steeper slope of the line), but the genetic difference is expressed also under poor conditions. Most research on genotype - environment interaction in dairy cattle supports this model (review by Syrstad, 1976). In studies on field data, progenies of various bulls have been found to rank very similarly over a wide range of production levels. The same was true when progenies of the same bull in Mexico and U.S./Canada were considered (McDowell <u>et al</u>., 1975). A recent review of dairy cattle crossbreeding in the tropics (Syrstad, 1989) suggests that the relative merits of two genetic groups (1/2 vs. 3/4 exotic inheritance) is independent of production level.

Figure 1c describes a situation in which the different responses of the two genotypes to environmental improvement results in reversed ranking. Genotype A is the better under good conditions, while B is superior when the environment falls below a given level. This might occur in cases when the environment varies over a very wide range. An example from beef cattle is presented by Hearnshaw & Barlow (1982). Crosses of Hereford with Friesian, Simmental and Brahman (American Zebu) were compared under good, intermediate and poor pasture conditions. The Friesian and Simmental crosses were the best on good pasture, while Brahman crosses were superior on poor pasture. In dairy cattle, Buvanendran & Petersen (1980) found almost no relationship between the performance of daughters of the same bull in Denmark and Sri Lanka. However, the number of daughters in Sri Lanka was small, and the lack of association might be incidental.

Model (b) suggests that the best breeding strategy is to select breeding animals on their performance in a good environment, as this is when the genetic differences between animals is most clearly expressed. But this would be dangerous if model (c) should be correct. The safest, and usually the most efficient, approach is to base selection on the merits of the animals as expressed under environmental conditions similar to those which their progenies will be exposed to.

<u>METHODS FOR GENETIC IMPROVEMENT</u>

a) <u>Selection within the local population</u>

Cattle indigenous to the tropics have, except in very few cases, been subjected to only little artificial selection for increased milk production. In view of the impressive results achieved by selection in many temperate dairy breeds there should be good prospects for improving the dairy potential of tropical cattle by the same method.

Genetic improvement per generation from selection depends on the variability of the traits considered, their heritability (i.e. the proportion of total variation which can be ascribed to genetic differences), and the intensity of selection. Variability, in terms of the coefficient of variation, is usually greater in tropical than in temperate cattle, but the variation in actual units is less. Studies of heritability based on sufficiently large amounts of data are few, but estimates reported fall within the same range as those from temperate countries. Intensity of selection is restricted by the reproductive rate, and is further reduced by early mortality, which often is high under tropical conditions.

Many dairy cattle breeding programmes claim a genetic improvement in milk yield of one to two percent per year. Of this improvement, 60 to 70 percent is derived from the selection of bulls on the basis of the performance of their daughters (progeny testing). This is achieved by a combination of accurate progeny testing (i.e. many daughters per bull) and intensive selection (many bulls tested per year). These conditions can be fulfilled only in large populations, comprising tens (if not hundreds) of thousands of females, artificial insemination, and widespread milk recording.
In most tropical countries such populations do not exist, and are not likely to be available in the foreseeable future. Instead a breeding programme might have to be established in a single herd or a few cooperating herds. In order to make progeny testing worthwhile, even strictly on genetic grounds, several hundred females would be needed. Still, the high costs involved might not make such a programme attractive. But if the herd serves as a nucleus herd, also providing bulls for breeding outside the herd, the benefit of genetic progress will in turn be transmitted to a much larger number of animals, and it might be justified to maximize genetic progress in spite of high costs. Thus a rather small breeding scheme can have tremendous impact if organized and operated properly.

b) Introduction of improved tropical breeds

Some breeds of tropical cattle, e.g. Sahiwal and Red Sindhi, have been selected for increased milk yield over a long time and have reached a much higher dairy potential than most cattle in the tropics. This is a genetic resource which should be exploited for upgrading of unimproved stock. After a few generations of back-crossing to bulls of the improved breed, the inheritance of the local cattle has been almost completely replaced by the improved inheritance. The risk of losing adaptability to local conditions by this method is small, a breed like Sahiwal has shown to adapt well to conditions in four different continents. An improvement which would require ten generations of intensive selection could be obtained in two or three generations of upgrading with an improved breed. Unfortunately the number of animals of improved tropical breeds is small, and breeding stock of high quality are not easily available.

c) Introduction of temperate breeds

Reports on the high milk yields in some temperate countries have spread the belief that the importation of European-type dairy breeds is the solution to the problem of low production levels in the tropics. In some cases introduction of temperate breeds has been successful but much more often the experience has been disappointing and sometimes almost disastrous. Diseases, high mortality rates and low fertility have been frequent problems among the imported animals and their progenies, and animals which have survived have failed to reach the expected production levels. Offspring born in the tropical country have often produced much less than their dams, which were imported as heifers. The lack of adaptation to tropical conditions has been obvious. On the basis of experience up to this time, purebred European-type dairy cattle can be recommended in the tropics only if climatic stress is moderate, health services are easily available and reasonably good feeding is practised.

d) Crossbreeding with European type cattle

Crossbreeding of tropical cattle with cattle of European-type breeds has occurred for more than one hundred years, and a large number of reports has been published. In most cases, females of local stock have been mated to bulls of the imported breed or by the use of imported semen.

In almost all cases, crossbreeding with a European breed led to a dramatic increase in milk yield in the first crossbred generation (F1), compared with the local stock. The crossbred females calved at a much younger age than native animals, produced two to three times more milk and had longer lactations, shorter dry periods and shorter calving intervals. Mortality and susceptibility to disease were only slightly higher than in native cattle.

These favourable results were, naturally, ascribed to the superiority of the exotic inheritance, and it was tempting to introduce more of it by backcrossing to exotic bulls. But the expected further improvement did often not occur and in many cases a decline in performance was observed. Problems of high mortality and reduced fertility increased as the level of exotic inheritance increased towards 100 per cent.

When it had been found that upgrading towards the European breed was not advisable under most conditions, the next step was to try to stabilize the level of exotic inheritance by mating F1 males and females together. But again the results were often disappointing. In almost all projects the performance of the second half-bred generation, F2, has been much below that of F1. Age at first calving and calving intervals have increased considerably and milk yield has dropped by up to 30 per cent.

A summary of results from 54 sets of data reported from crossbreeding experiments in various regions of the tropics is presented in Table 1. The good performance of the first crossbred generation (F1) and the deterioration in the next generation (F2) are clearly demonstrated. The most obvious explanation is the presence of hybrid vigour (heterosis); this effect is maximized in the F1 but half of it is expected to disappear in the F2 and forward generations. In addition other genetic mechanisms might also be involved.

The great effect of hybrid vigour in crosses of zebu x European-type cattle might be expected because of the wide genetic distance between the two types (Cunningham and Syrstad, 1987). Furthermore it has been suggested that hybrid vigour is more important under stressful than under favourable environmental conditions (review by Barlow, 1981). The breeding strategy for dairy cattle in the tropics should therefore also aim at exploiting hybrid vigour. Exactly how this can be done under various conditions is still a question for discussion, and more research is needed.

Table 1. Performance of zebu cattle, European type cattle, and their crosses in the tropics. Summary of 54 sets of data. Source: Syrstad (1988).

Proportion European cattle	Age at first calving, months	Milk yield, kg	Calving interval, days
0 (i.e. zebu)	43.6	1052	459
1/8	40.1	1371	450
1/	37.5	1310	435
3/8	36.1	1553	435
1/2 (F1)	32.4	2039	429
5/8	33.8	1984	432
3/4	33,9	2091	450
7/8	34.4	2086	459
1 (i.e. European)	31.6	2162	460
1/2 (F2, from F1xF1)	33.7	1523	449

Table 2. Comparison of F1 and backcrosses (1/2 and 3/4 European inheritance) at low, intermediate and high production levels. Summary of 30 sets of data. Source: Syrstad (1989).

Production level	Average milk yield, kg F1	Backcrosses
Low (<2000 kg)	1487	1605
Intermediate (2000-2405 kg)	2175	2218
High (>2405 kg)	2798	2698

REFERENCES

Barlow, R. 1981 Experimental evidence for interaction between heterosis and environment in animals. Animal Breeding Abstracts 49: 715-737.

Buvanendran, V. & Petersen, P.H. 1980 Genotype-environment interaction in milk production under Sri Lanka and Danish conditions. Acta Agriculturae Scandinavica 30: 369-372.

Cunningham, E.P. & Syrstad, O. 1987 Crossbreeding *Bos indicus* and *Bos taurus* for milk production in the tropics. FAO Animal Production and Health Paper 68, 90 pp.

Hearnshaw, H. and Barlow, R. 1982 Genotype x environment interaction in crosses among beef cattle. Proceedings of the World Congress on Sheep and Beef Cattle Breeding, Palmerston North, New Zealand, 1980, Vol. I: 469-474.

McDowell, R.E., Wiggans, G.R., Camoens, J.K., Van Vleck, L.D.and St.Louis, D.G. 1975 Sire comparisons for Holsteins in Mexico versus the United States and Canada. Journal of Dairy Science 59: 298-304.

Syrstad, O. 1976 Interaction between genotype and nutrition in dairy production, - a review. World Review of Animal Production 12: 33-38.

Syrstad, O. Crossbreeding for increased milk production in the
1988 tropics. Norwegian Journal of Agricultural Sciences 2: 179-185.

Syrstad, O. Dairy cattle crossbreeding in the tropics: The importance
1989 of genotype x environment interaction. Submitted for publication in Livestock Production Science.

MATCHING LIVESTOCK SYSTEMS WITH AVAILABLE RESOURCES

by

T R Preston

DEVELOPMENT MODELS

The Third World

In the tropical regions of the Third World in general and specifically at the level of the small farm, livestock production is in crisis. This crisis is closely related with the production models which were implanted in Third World countries during the post-war period and were intensified in the decades of the 60's and 70's. In order to introduce these new technologies, many Third World countries established credit mechanisms, agricultural research programmes, rural extension schemes, and training programmes aimed at increasing production by promoting an increase in the productivity of crops and animals.

The model that was advocated for the agronomic sector aimed to bring about a "green revolution", specifically in cereal grain production, that would solve the nutritional problems and hunger suffered by millions of people in the Third World. "Improved" production systems were promoted, based on high performance germ plasm, monocultural practices and the intensive use of capital, machinery, and costly imported inputs such as fertilizers, pesticides and herbicides.

In the livestock field, priority was given almost exclusively to the introduction of systems, the technical and economic bases of which were derived from experiences in the temperate, "industrialised" countries, where emphasis had been on low labour inputs, high use of capital, and intensive specialised production methods aimed at market expansion. For example:

- Cattle production systems were based on the American and Australian models that employed extensive grazing on pastures established in regions previously in natural forest. The result has been an alarming increase in erosion and destruction of ecosystems and watersheds.

- Pig, poultry and milk production systems were based on "economy of scale", involving an ever increasing dependence on "imported " inputs (feed grains and protein meals, germ plasm, drugs, equipment and fossil-derived fuel), and an overall negative effect on employment opportunities, especially in rural areas.

The result of these activities has been an increasing dependence on imported inputs, increased costs of production, reduced rural employment, contamination of the environment and destruction of ecosystems, deforestation and under-utilization of available resources.

The industrialised countries

The transfer of livestock technologies from industrialised to developing countries, has obviously been largely unsuccessful. Quite apart from the reasons for such failures, it is relevant to question the basic concepts governing the models currently employed in the industrialised countries. For, contrary to what is so often assumed, it may not be desirable - even if it were, technically and economically feasible - to attempt to achieve in Third World countries the styles and standards of living currently "enjoyed" (??) in the industrialised countries. Leaving aside the social issues, an assessment of the present agricultural situation in most industrialised countries shows that:

- agricultural products - especially those of animal origin - are expensive to produce and to buy,
- present production systems are wasteful and cause considerable ecological damage,
- the systems of production and the products that are produced are frequently associated with stress both for animals and humans, and
- the dependence on, and excessive use of, fossil fuel based inputs is causing an alarming increase in atmospheric carbon dioxide concentrations, which is the main contributer to the warming of the earth's atmosphere - the "greenhouse effect."

It hardly seems sensible to encourage the developing countries to commit their scarce economic resources to livestock production programmes which may eventually arrive at the same inappropriate end-point.

Eco-development and self-reliance

The concept of eco-development has been proposed by Third World economists as an alternative to the classical development models derived from the industrialised countries, which have proved to be unsustainable when introduced into developing countries. The basic feature of eco-development is that the means of improving the quality of life of a community should be sought within a framework bounded by the limitations - environmental, social and economic - governing the activities of the community. The means of achieving such aims should be determined by the principles of self-reliance; in other words, the technologies used should be decided and executed by the community and should not be dependent on outside events and forces.

DESIGN OF LIVESTOCK TECHNOLOGIES FOR THE TROPICAL THIRD WORLD

Goals and means

Improved technologies are essential tools in all forms of development. Past mistakes in technology transfer can be traced to the failure to understand the fundamental issues which must be considered before embarking on the design of technologies. Of these, the major one is: what are the constraints governing the design of the technologies? Are these the same for both developed and developing countries or are there basic differences that should be taken into account? Experience from the Third World tells us that while the scientific principles which underly technologies are the same, the technologies themselves are likely to be quite different. Some of the reasons for this statement are set out in Table 1.

These differences help to explain why there are conflicts concerning the strategies that should be applied when national governments, supported by international and bilateral technical assistance agencies (whose policies are largely determined by professionals from the industrialised countries), attempt to introduce innovations in the field of livestock research, technology transfer and training.

Table 1: Design of livestock technologies for industrialised and developing countries; goals and means

	Industrialised	Developing
Climate	Mostly temperate	Mostly tropical
Role of livestock	Specialised	Multipurpose
Target group	The rich	The poor
Resource base:		
Feed	Starch-protein	Fibre-sugars
Genetic	Improved	Native
Capital	High	Low
Labour	Low	High
Mechanisation	High	Low
Agrochemical	High	Low
Infrastructure	Good	Poor
Marketing	Good	Poor

Experience has shown quite clearly that it is counter-productive in developing countries to base these activities on the models that have been, or are currently being, applied in the industrialised countries. The conclusions of a recent evaluation report on Dutch assistance to the livestock sector in Third World countries (Netherlands Development Cooperation, 1987) reveal how much time and effort have been wasted in these endeavours:

> ..."The animals (Dutch dairy cattle) were generally unable to adjust to local conditions; climate, feed and management systems all posed problems. The cattle were unsuited to small farmers' needs; they could not be used as draught animals and often suffered from disease stress, leg problems and infertility" ....
>
> ...."the experience of twenty years revealed that the route (intensive livestock projects) had been ill-chosen".

THE "GREENHOUSE EFFECT"

Discussing the issues of the world food crisis, population growth and renewable energy, Dumont (1989) stated:

> "..Now an even more formidable threat has appeared on the horizon. Until recently, all forecasts (of crop yields) were based on the virtually certain knowledge that the world's various climates were invulnerable to major meteorological upheavals. That last lingering certainty has passed.....
>
> ....it is now established beyond the slightest doubt that increased carbon dioxide levels in the air as a result of an excessively rapid increase in the use of fossil fuels, among other things, has caused the world's climate to warm up since the beginning of the 70s.
>
> ....The warming-up process will raise the level of the sea and threaten the existence of all the great river deltas, especially those in Asia..
>
> ...Increased temperatures have aggravated droughts and thus reduced the flow of the world's great rivers...
>
> ...Since the last ice age 18,000 years ago, the world temperature has risen by only four degrees. It is now feared that by the year 2050, only 61 years from now, the temperature increase could be between two and six degrees."

TOWARDS A NEW STRATEGY FOR SUSTAINABLE RURAL DEVELOPMENT

Matching production systems with available resources

The situation that has been described makes it obligatory to reorientate present production systems and to develop a new agricultural policy based on optimal use (instead of misuse) of the earth's natural resources. Such a policy must be rural, rather than urban, orientated and a notable feature of it will be the economic strengthening and increased independence of the small-scale farmer.

It has been proposed (Preston and Leng, 1987) that such a strategy be based on the development of agricultural systems which integrate production of food, fuel and fertilisers, and diverse livestock species, with emphasis on the utilization of:

- Existing under-utilised local resources and wastes (e.g., crop residues, livestock excreta and agroindustrial byproducts)

- New resources derived from more efficient agronomic systems based on improved utilization of solar energy, soil, water, genetic diversity and people - which are the natural resources of the tropics. Specific reference is made here to the use of sugar cane and forage trees and shrubs as the principal elements in such a scheme.

In the long and short term, it is hypothesised that the alternative to fossil fuel is in biomass derived from solar energy capture and that this is more viable and desirable than energy from nuclear sources, especially when environmental and social issues are taken into account. Furthermore, it is proposed that there need be no conflict, indeed the prospects are for complementarity, in the use of biomass to satisfy both food and fuel needs.

Such a policy presupposes a series of conditions, principal among which are the following:

- crops and cropping systems must be chosen which permit maximum capture of solar energy and its conversion into biomass;

- optimum fixation of atmospheric nitrogen in relation to the nutrient needs of the selected crops and associated livestock systems;

- fractionation of the crops to satisfy dual needs of food/feed and fuel;

- the livestock components of the system should address the complementary needs of monogastric and herbivorous animal species.

- The overall system should:

    - be at least self-sufficient in, and preferably a net exporter of, energy,
    - not contaminate the environment,
    - not destroy natural ecosystems,
    - optimise employment opportunities, and
    - promote a maximum degree of self-reliance.

MEASURES THAT CAN BE TAKEN TO REDUCE EMISSIONS OF CARBON DIOXIDE AND METHANE

To the general policy described above, there must now be added a series of additional recommendations in order to address the specific problem of the "greenhouse effect". The following measures can be expected to lead to reduced emissions of methane and carbon dioxide. Not all are immediately executable, but they indicate what should be the long term goals in order to attain and maintain a balance between sources and sinks of carbon dioxide and methane.

- Giving priority to the growing of crops which are most efficient in fixing carbon dioxide into biomass (eg: perennial tropical forage crops and trees).

- Encouraging agricultural production in the tropics of the Third World, instead of promoting self sufficiency in industrialised temperate countries. This is because food production systems in industrialised countries are highly dependent on fossil fuel-derived inputs. By contrast, most Third World tropical country systems employ animal and human power, rather than machinery, and they have a much greater potential for developing biomass-derived fuels. Means to this end would be the elimination of tariffs on food imports from the tropics, and of subsidies to farmers in the industrialised countries, and by applying an environmental tax to the use of fossil fuel since this is the main cause of the "greenhouse effect".

- In the tropical countries, wetland rice should be discouraged and more emphasis given to dryland cereal production for human consumption. Cereal growing for animal consumption in tropical countries should be actively discouraged, and emphasis given to perennial forage crops and forage trees as the basis of intensive animal production.

- Grazing systems in the tropics should be actively discouraged, in favour of complete or semi-confinement of animals. This will permit planting of existing grazing lands with forests (especially multipurpose forage trees) and favour the greater use of crop residues as animal feed (instead of burning them).

- A massive programme is needed to promote strategic supplementation of ruminant diets in Third World countries in order to optimise rumen function (which leads to reduced methane and CO_2 production). This will also lead to increased production of food, or the keeping of fewer animals.

- Non-ruminant species (especially pigs, poultry, rabbits) should be favoured over ruminants as meat producers, since they produce less methane and carbon dioxide per unit of product.

- Low-cost biogas digestors must be an essential element in all units where livestock are confined.

- Human organic food waste must be recycled through pigs, and/or earth worms, instead of being allowed to ferment in land fills (giving rise to methane) or to be incinerated (producing carbon dioxide).

- Maximum support should be given to research and development efforts which will enable fossil-derived fuels to be replaced by biomass-derived fuels. Gasification of biomass to produce hydrogen and carbon monoxide (can be used directly as fuel or as substrates for chemical industry) would appear to be the most appropriate technology to promote.

REQUIRED INFRASTRUCTURE

Political or Technological Reform?

It is usually assumed that the first constraint to rural development is the need for reform of land tenure. However, it is becoming increasingly apparent that, with or without agrarian reform, there is no way that farming systems can absorb the "landless" labour force that exists in rural areas in most of the Third World. New solutions are needed and these must be based on proposals for technological as well as political change.

Tropical countries offer exciting possibilities for technological reform, because of the largely untapped potential for biomass production in regions blessed with abundant supplies of solar energy, high mean temperatures and rainfall. However, the realization of such potential wil require an original approach not only to the growing of the biomass but also its utilization.

Energy (with food as a byproduct) will be the key to such schemes, and the utilization of biomass as feedstock for a chemical industry will be as important as providing a substitute for present fossil-based liquid fuels. Rural industries based on "biomass refineries" promise to solve problems which are immediate, such as increasing rural employment, and of longer term, as is the prospect of developing a viable and safe alternative to non-renewable energy sources, both fossil and nuclear.

The need to reverse the greenhouse effect, coupled with the concern about the risks and the environmental contamination associated with the nuclear option, is a golden opportunity for the tropical regions to exploit their largely untapped resources inherent in the opportunity to use solar energy throughout the year. Existing rates of photosynthesis permit the capture of 10 times more energy than is presently consumed as fossil fuel (Hall, 1984: personal communication). This is being achieved with an overall global efficiency of only 0.2%. By contrast, a perennial tropical crop such as sugar cane fixes solar energy at 10 times this rate (2% annually) (Bassham, 1978). Tropical trees are almost as efficient and most have the added virtue of being able to fix ambient nitrogen in their root system.

STRATEGIES FOR MILK PRODUCTION SYSTEMS IN THE TROPICS

Where, in the above scheme, does tropical milk production fit and what strategy should be followed in establishing this kind of activity? The starting point must be an analysis of the actually and potentially available resources. There are no specialised tropical dairy animal breeds, other than the Riverine type of bufffalo. Furthermore attempts to create them have not proved to be sustainable. The impact of the Australian Milking Zebu and of the Jamaica Hope, for example, has been minimal outside the immediate areas where they were developed.

The immediately available and numerically important cattle resources in the tropics are *Bos taurus* beef cattle. There are many advantages from using these as a basis for milk production through crossbreeding, foremost among which is the increase in productivity and in biological efficiency that results when milk and beef production are combined in the same animal. Converting existing extensive beef cattle systems in tropical countries into dual purpose milk-beef enterprises will increase their productivity and biological and economic efficiency. In ecological terms, this means a global reduction in methane production per unit of animal product, and the possibility of reducing total animal numbers (fewer, more productive animals consuming the same basic feed resources).

From the nutritional standpoint, the increase in productivity required in a dual purpose, as opposed to a specialised beef animal, does not entail substitution of local feed resources by exotic (usually grain-based) "balanced dairy feeds" as is the case when specialised dairy breeds and systems are introduced (Netherlands Development Corporation 1987). What is needed is strategic supplementation with rumen activators and bypass nutrients, which can be met by judicious use of mostly locally available tropical agroindustrial byproducts (e.g., multi-nutritional blocks from molasses and urea, rumen micro-nutrients and bypass macro-nutrients from oilseed cakes and tree foliages (see paper by Leng, this conference)).

CONCLUSIONS

A new era is dawning in development strategy. Participation is mandatory in the setting of goals and identification of means. It is also becoming apparent that, because both the goals and the means are not the same, the initiatives taken by developing countries in establishing their own development strategies must not only be respected, but may also serve as stimuli for more effective cooperation between developed and developing countries.

Inadequate human nutrition is still the most immediate problem in most developing countries. But it is now being realised that the solution is not simply to increase productivity but to tackle more fundamental issues, foremost among which is the warming of the earth's atmosphere, caused by increased ambient concentration of carbon dioxide (mainly due to increased use of fossil fuels). This threatens, in the shorter term, to reduce crop yields and, in the longer term, heralds unmitigated disaster through flooding of river deltas.

Reversing the "greenhouse effect" will require promotion of solar efficient perennial crops, and forage trees, which simulate forest ecosystems and provide a sink for carbon dioxide. The biomass from these crops should lend itself to fractionation into low and high fibre components, the former being the basis of intensive confinement production of monogastric animals, while the latter can be converted into versatile energy-yielding substrates suitable either as ruminant feeds or, through the process of gasification, as the basis of a chemical industry (hydrogen and carbon monoxide). Liquid and solid wastes can be recycled through biodigesters and earthworms with much reduced emissions of methane and carbon dioxide, compared with processing them through conventional oxidation lagoons and land fills.

In the field of livestock production, increasing emphasis must be given to systems which reduce methane emissions per unit of livestock product, at the same time permitting greater use of locally and potentially available resources. In this respect the two major approaches are: giving greater emphasis to monogastric species (especially pigs) as meat producers, and adapting presently inefficient extensive beef operations into dual purpose milk-beef systems.

These technologies, which are now being developed in Third World countries, will lead to more sustainable systems of livestock production, to employment generation, to increased availability of renewable energy and - most importantly -to a reversing of the "greenhouse effect". The challenge facing governments of industrialised and Third World countries alike is to be able to accept that development without either fossil or nuclear fuels is not only technically feasible but will bring with it much needed sociological and ecological benefits through the greater role that will be given to rural areas as the future source of both feed and fuel.

The implementation of these new strategies will require a greater appreciation of:

- Communication as a means of promoting:
    - understanding of changing priorities,
    - awareness of common problems and the means of overcoming these, and
- Relationships founded on technological support rather than economic dominance.
- The concepts of ecodevelopment and self-reliance when designing and implementing technologies.

REFERENCES

Bassham, J.A. Photosynthetic productivity of tropical and temperate
1978 crops. Caribbean Consultancy on Energy and Agriculture. Santo Domingo, Dominican Republic. November 29th, 1978.

Dumont, R. Education of women offers best hope for world food crisis.
1989 *Le Monde*: Paris, March 30. pp. 16-17.

Netherlands Development Cooperation. Assisting livestock development.
1987 Evaluation Report. Netherlands Development Cooperation (Ministry of Foreign Affaires): The Hague.

Preston, T.R. and Leng, R.A. Matching ruminant production systems with
1987 available resources in the tropics and sub-tropics. Penambul Books: Armidale, Australia pp. 245.

Preston, T.R. and Murgueitio, E. Tree and shrub legumes as protein
1987 sources for livestock. In: Forage legumes and other local protein sources as substitutes for imported protein meals (Editor: D. Walmsley). CTA, Wageningen and CARDI, Trinidad pp. 94-104.

NUTRITIONAL CHARACTERISTICS OF TROPICAL FEED RESOURCES: NATURAL AND IMPROVED GRASSLANDS, CROP RESIDUES AND AGRO-INDUSTRIAL BY-PRODUCTS

by

M. Chenost and R. Sansoucy

INTRODUCTION

Numerous studies and reviews have already been completed on this topic, e.g. Osbourn (1976), Minson (1976), Stobbs (1976), Balch(1977), Chenost and Meyer (1977), Jarrige (1979), Göhl (1981), Lane (1981), Preston (1982) and Devendra (1988). In the context of the present consultation, we will therefore restrict ourselves to reviewing the main characteristics of the tropical feed resources which should be taken into consideration when defining diets and feeding systems in accordance with the new principles of ruminant digestive physiology and nutrition.

RUMINANT INTAKE AND DIGESTION

The individual cow's daily production depends not only on its genetic characteristics and its stage of lactation but also a great deal on the quantity and quality of nutrients to its intermediary metabolism. This supply is the result of the voluntary intake and the nutrient density of feed intake.

Voluntary intake depends both on:

- the appetite of the animal which varies according to the animal itself (age, physiological stage, former nutritional status, etc.) and to the environmental conditions (temperature, humidity, etc.) under which the animal is kept, and
- the specific characteristics of the feed.

The voluntary intake of feed depends essentially on the rate of degradation of its digestible matter into particles of a size small enough to enable their passage from the reticulo-rumen to the lower gut. This degradation is achieved by means of the chewing process (eating and rumination) and the microbial fermentation which takes place in the reticulo-rumen. The cell wall content and the magnitude and nature of lignification of these cell walls are amongst the most important factors which govern the degradability and the rate of passage of a forage.

Good microbial activity will require:

- adequate nutrition of the rumen microorganisms: energy in the form of ATP released from soluble and structural carbohydrates of the plant, thanks to the anaerobic fermentation; nitrogen in the form of ammonia generated by the hydrolysis of the fermentable nitrogen; minerals and vitamins;

- good chemical and physico-chemical rumen environment: pH (which should be as constant as possible and not below 6.5 to favour the cellulolytic microorganisms) and a regular outflow from the rumen. These conditions are not only dependent on the properties of the feeds but also on their rationing (number and frequency of meals, physical form of their presentation).

Nutrients required at the tissue level for both maintenance and milk synthesis are supplied by the end products of rumen fermentation (amongst which are the volatile fatty acids (VFAs) and microbial cell proteins) and by the dietary nutrients which have escaped rumen degradation and are digested in the intestine. Depending on the level of production of the host animal, it may be necessary to provide, in addition to the forage, dietary supplements in order to meet its nutritional requirements. These supplements should be administered in a certain amount and should possess characteristics, such that the rumen ecosystem is not impaired and generates the proper amount and relative proportions of microbial protein, VFA energy and glucogenic energy.

In order to define an optimum diet it will therefore be necessary to choose the feeds according to the quality and quantity of energy and nitrogen available. These characteristics cannot be determined by the classical routine analysis. In addition to Crude Protein content and Organic Matter (or energy) digestibility it is important to know:

- an estimate of intake, more particularly for those feeds which compose the basic part of the diet. A good indicator is the rate of their dry matter degradability in the rumen. This can be approached through the nylon bag "in sacco" technique which uses rumen-fistulated animals;

- an estimate of the respective parts of rumen degradable and undegradable ("by-pass") proteins available, respectively, for the rumen microorganisms and the host animal. This can be also approached with the nylon bag technique which can give an estimate of the extent and rate of protein fermentability. The French PDI (Protein Digestible in the Intestine) system (INRA, 1988) can distinguish between the various parts which are finally digested in the intestine, i.e. PDIA (Protein Digestible in the Intestine from dietary origin), microbial protein allowed by available fermentable N (PDIMN) and microbial protein allowed by available fermentable energy (PDIME). The sum of PDIA + PDIMN on the one hand, and the sum of PDIA + PDIME on the other hand give, respectively, the PDIN and PDIE values of a feed. Balancing a diet by supplementing the basic feed with the appropriate supplementary feeds is achieved when PDIN and PDIE values of the diet are equal and meet the production requirements. This system is being adapted, for instance, to the Caribbean feedstuffs (Xandé and Trujillo, 1985).

- Data on energy sources: rate (see above) and type of fermentation. The slowly fermentable energy released from the structural carbohydrates or the more easily fermentable energy released from high digestible cell walls (e.g. citrus, beet or fruit pulps, which both favour a cellulolytic ecosystem. The fast degradable carbohydrates of "sugar type" (molasses) or of "starch type" (cereals, roots and tubers, banana) which both hamper the cellulolytic ecosystem (drop in pH). The end products of the former are essentially C2 whereas those of the second favour C3 VFA. Other important information, but of course difficult to predict, is the good timing of the release of NH_3 and ATP for optimum microbial nutrition and thus synthesis and microbial activity. Finally, an assessment of the undegraded part of energy in the rumen (e.g. rice polishings or maize, rather than wheat or cassava) usable in the intestine for the tissue requirements (glucogenic function) is also important (Preston, 1982; Van Es, 1985).

All these considerations are undoubtedly more important in tropical than in temperate regions even if the levels of animal production are lower. In fact, shortages of nitrogen (tropical feeds also contain less by-pass protein) and of digestible cell-wall energy may occur quite often in tropical countries. It is therefore important to be able to choose the proper missing components among the other locally available resources. As already discussed in several instances, supplements of the tropical basic diets have often more than an additive effect on both intake and animal performances (Preston, 1982; Van Es and Taminga, 1987).

We will briefly distinguish between the basic resources which compose the main parts of the diet and the other various resources which can supplement them.

The former are pastures and green fodders which are of course the principal natural ruminant feed. They are also crop residues, including the fibrous agricultural residues (FAR) which can be used as a substitute (partly or entirely) for herbage in those populated regions like South East Asia where land must firstly be devoted to production of food for man. Another group is the perennial food crops (e.g. sugarcane, bananas), and also roots and tubers which were formerly grown for man and are now more and more considered as feed, either for the dry season or even as the basis of the diet for new feeding systems.

Table 1. Main nutritional characteristics of the principal categories of tropical feed resources.

Feeds	Rumen fermentation		Observations
	Energy	Nitrogen	Utilization
Pastures Green fodders Forages	slowly fermentable VFA - C_2	fair CP content fermentable	-fair intake -lignified -rumen function and rumination Basis of diet
Crop residues: straws stovers canetops	slowly fermentable	very low CP content	-increased OMD and intake with treatment -need fermentable N + PDIA Basis of diet
Feed crops: sugarcane (whole)	slowly) fermentable quickly)	low CP content	-need fermentable + PDIA nitrogen Basis/supplement
Foliages - tree crops including Leucaena, Glyricidia, etc.	fermentable C_2 - C_3	very high CP unfermentable* (by-pass)	-rumen function -aa sources(PDIA) -good intake Supplement/Basis
Agro-industrial byproducts:			
<u>energy</u>			
molasses	easily fermentable ("sugar" type)	low N	Supplement/Basis
pulps (citrus)	easily fermentable ("cell wall" type)	low N	
<u>energy + N</u>			
bran/polishings	fermentable + lipids LCFA	unfermentable N source	Supplement -PDIA + bypass energy,glucogenic
<u>nitrogen</u>			
oil cakes + seeds animal/fish	 lipids	high CP PDIA a.a.	 Supplement
NH_3 - urea		fermentable	industrial NPN

*Further research is needed regarding the tannin effect on digestion (enzymatic) in the intestine

The other category is roughly made-up of the agro-industrial and various by-products which can be utilized only as part of the diets.

We will now consider the way these feed resources can be utilized in the appropriate combination so that:

- intake of the basic components is maximized,
- animal performances are optimized, and
- cost of diets is minimized.

Although pastures and green fodders are the principal natural feeds for ruminants, there are also other feed resources which can be used as substitutes during the dry season (e.g., crop presidues), supplements (agro-industrial by-products: cereal brans, molasses, oilcakes, etc.) or as the basis of the diet (sugar cane, roots and tubers, bananas).

GREEN FODDERS, HERBAGES AND PASTURES

It is well recognized that the tropical herbaceous and shrub plants become high in lignified carbohydrates and low in total nitrogen when they mature. In addition their mineral content is low and unbalanced; phosphorous is amongst the most frequent deficient macro-elements.

The digestibility of tropical forages decreases at a lower rate than that of temperate ones but this decrease starts earlier and from a lower value at the young vegetation stage (Chenost, 1975; Evans, 1977). As a result, tropical grasses, and to a lesser extent legumes, always have a lower digestibility than the temperate ones (Minson, 1976) as shown in Table 2. In fact, the high content and the type of encrustation of lignin in the plant tissues and cell walls and the low N supply to rumen microbes are reasons which lead to a slow rate of breakdown and passage of particles to the lower gut and reduced intake of tropical grasses.

However, except in the case of natural pastures in dry tropical areas, tropical pastures have a tremendously high dry matter productivity. This productivity enables maintenance of high stocking rates (carrying capacity) as shown in Table 3.

The yields of tropical C4 grasses (e.g. *Digitaria decumbens*) responds linearly to annual rainfall (or water supply when irrigated) when fertilized with nitrogen up to 400 kg N/ha (Salette, 1970). Nitrogen fertilization however does not increase the animal's daily production since it has very little or no effect on digestibility and voluntary intake.

Milk production per area unit can thus reach high levels, thanks to the stocking rate, but with low individual production per animal,

Table 2. Examples of digestibility, intake and rumen fermentation by sheep of some tropical grass hays. (from O. Kawamura *et al.*, 1985)

Nature of grasses	Regrowth number (*)	Vegetative stage	Dry matter intake g/kg LW 0.75	Dry matter digestibility (per cent)	Crude Protein content (per cent DM)	ADF	VFA mm/100ml	C2 percent	C3 VFA
Green panic	1	flowering	47.7	58.9	10.2				
	2	heading	45.9	53.4	13.5				
	3	heading	52.9	54.4	11.9				
Sudan grass	1	flowering	42.0	53.7	11.9				
	2	vegetative	50.7	55.0	15.6	37.5	5.62	75.1	13.0
	3	vegetative	48.0	58.0	14.9	to	to	to	to
						49.1	8.96	82.6	18.2
Rhodes grass	1	vegetative	44.2	56.3	10.0				
	2	vegetative	42.4	60.7	13.2				
	3	flowering	47.4	59.6	12.2				
African millet	1	vegetative	49.5	60.7	14.0		10.24		
Italian Ryegrass	1	heading	70.1	72.5	13.6	28.1	12.24	70.1	22.3
	2	heading	59.1	58.6	11.1	41.8	10.20	71.6	21.0

(*) wilted and air dried

ADF = Acid Detergent Fibre; VFA = Volatile Fatty Acids; C2 = Acetic Acid; C3 = Propionic Acid.

as shown by numerous authors quoted by Evans (1977) (Table 4). Such low daily milk production levels (seldom higher than 12 kg) may hamper the duration of the lactation curve.

Many studies have shown that trying to increase individual production by exploiting the grass cover at an earlier stage of growth is wasteful and uneconomic. A recent study on buffaloes has also lead to the same conclusions (Wanapat and Topark-Ngarm, 1985). In fact, the loss in DM production/ha is not compensated for by the very small benefit, in terms of DOM intake, which could be expected from a faster turn-over.

Whereas a 4,500 kg milk lactation needs, in addition to a typical temperate forage-based diet, an average of 150 g concentrate for each kg of milk produced, the same level of milk production requires an average of 300 to 350 g concentrate in the case of a typical tropical forage-based diet. Such amounts are uneconomic (except when concentrates are subsidized) and illogical (substitution effect of concentrate depresses the DM intake of forage). It is therefore necessary to resort either to an improvement of the basic diet or to design a strategy of supplementation taking into account other local feed resources (see below).

In areas where rainfall is higher than 750 mm per year, it is possible to oversow natural pastures with legumes which, most of the time, are not present in the primary grass cover. The effect of legumes is two-fold: firstly fixation of substantial amounts of N and therefore increase of the production of the associated grasses, and secondly an increase in the feeding value of the grass cover resulting from the higher N content and by-pass protein, the higher OMD and intake of legumes. A lot of research work has shown the importance of fodder legumes (Table 4) on the individual cow's daily production. The strategy to be finally adopted regarding the type of pasture (pure grasses versus grass/legume association) is however not only a nutritional problem but also an agronomical and managerial one.

But tropical green fodder, which represents the cheapest sources of forage, cannot in general ensure high individual milk secretion levels. The main limiting factors are intake and N content and quality.

CROP RESIDUES

The fibrous agricultural residues (FAR) represent a considerable potential forage resource in the populated countries where land must be devoted to human food production as a priority. A comprehensive review of their potential in the developing countries and of the strategies for expanding their utilization has been achieved respectively by FAO (1985) and IDRC and ICAR (1988).

Table 3. Carrying capacity and milk production per hectare from various pasture systems (from Stobbs, 1976, quoted by Jarrige, 1979).

Pasture system	Stocking rate (cows/ha)	Milk production (kg/ha/year)
- unfertilized grass	0.8 - 1.5	1,000 - 2,500
- grass-legume	1.3 - 2.5	3,000 - 8,000
- nitrogen fertilized grass (+P, S, K)	2.5 - 5.0	4,500 - 9,500
- nitrogen fertilized grass, irrigated (+P, S, K)	6.9 - 9.9	15,000 - 22,000

Table 4. Milk production from tropical pastures without supplementary feed (extracted from Evans, 1977)

Pasture	Stocking rate (cow/ha)	Breed	Milk yield kg/cow/day	Milk yield kg/ha/yr
Unfertilized pastures				
P. maximum *M. miniflora*	1.1	Jersey	6.8	2,667
P. maximum	1.0	Jersey/Criollo	6.9	2,667
D. decumbens	1.5	Friesian/Zebu	6.9	3,760
Grass-legume fertilized pasture				
P. maximum/ Glycine	1.3-2.5	Friesian	12.4-13.7	4,954-8,221
D. decumbens/ Centro	1.7	Friesian/Zebu	7.3	4,530
Nitrogen fertilized pure grass				
D. decumbens	2.5	Jersey	6.8	6,014
D. decumbens	6.9	Friesian/Zebu	10.9	17,408
D. decumbens	8.0	Jersey	6.5	22,466

Amongst the world's total crop residues maize yields the largest amount and wheat, rice and paddy and pulses each yield about half the amount of maize. The remainder consists of sorghum stovers, barley straws, sugarcane tops and leaves, roots and tubers, oil plants stovers and foliage (Kossila, 1985). They are still underutilized as feed resources, except in Asia where they form the first component of the ruminants' diet.

Their feeding value is limited by their poor voluntary intake, low digestibility and low nitrogen, mineral and vitamin content. In addition they are very slowly fermented in the rumen. In fact, they consist essentially of lignified structural carbohydrates, since they represent the dead aerial part of the mature plant after harvest.

The use of FAR as cattle feed has generated considerable research work in the last 20 years but unfortunately much less development application. However, they can represent the basic part of ruminants' diet provided:

- conditions for their good cellulolysis are met (rumen activity), and
- additional nutrients required for productive functions (host animal), e.g. PDIA and energy escaping rumen fermentation are properly supplied (do not impair the above condition as stated in para. II).

Their better digestive utilization can be achieved either through an appropriate supplementation (legumes, molasses, fruit pulps, poultry manure, urea, etc.) or chemical pre-treatments (urea/ammonia treatments) which both facilitate the microbial breakdown of the cell-walls. Appropriate supplements which enable a good cellulolysis can be chosen among the local feed sources on the ground of the characteristics listed in Table 1: the appropriate fermentable N supply can be of natural (poultry manure) or industrial (urea) origin; the fermentable energy (of the "digestible cell-wall" type) is typically fresh grass or good quality foliage and, of course, all the easy digestible agro-industrial pulps, e.g. citrus,pineapple, etc. The breakdown of FAR can also be improved by chemical treatments (Sundstøl and Owen, 1984) among which urea-generated NH_3 is probably the technique which best fits in with the socio-economical conditions found in tropical developing countries where inputs must be kept at the lowest level possible.

Treated or not, FAR must be combined with feed supplements which provide adequate nutrients to the rumen microorganisms. The former, however, still require to be known with a better accuracy than at present. Recent research works (Silva and Ørskov, 1988; Ramihone, 1987) have shown the importance - in addition to NH_3-N - of true protein sources on the cellulolytic microbial growth. This is another reason for supplying any cheap protected protein source (e.g. legume trees and foliages, ricebran), which, as seen above, are necessary for the production requirements of the host animal.

FOOD CROPS

Various perennial food crops which were formerly grown only for human consumption are now more and more considered as feed sources for the dry season and even as the basis of feeding systems. The main ones are sugarcane, cassava, banana, etc.

Amongst the various reasons for such their increasing use, two are probably most important. The first is the tremendous dry matter productivity/ha of these crops. The second one lies in the fact that, as opposed to conventional fodder crops, their nutritive value is not affected by the age of the plant which has already reached its stage of maturity. Their exploitation is therefore very flexible and easier than that of herbage.

1) The most typical example is probably the sugarcane, exploited as a whole plant. Sugarcane could play the same role in tropical animal production as the forage maize - whole plant - in the temperate countries. First considered in experiments by Preston in the 1970's, the sugarcane (whole plant) is typically the addition of two opposite types of forage components: structural carbohydrates of low and slowly digestible energy and soluble carbohydrates (sucrose) rapidly fermentable. In addition its N content is very low.

Whole sugarcane-based systems have proven to be technically and economically a very attractive solution for small to average dairy or dual purpose units in sugarcane producing countries where areas for fodder pastures are limited. As described by Preston and Leng (1978) the deficient nutrients may essentially be provided by locally grown (or available) feed resources:

- fermentable N (PDIN) by green fodder or crop foliages and leaves (e.g. those of cassava) or by urea,(industrial NPN source);
- "by-pass energy" by rice polishings and/or roots, tubers (cassava) and fruit (bananas - banana rejects);
- unfermentable nitrogen ("PDIA"/"by-pass N") by legume-trees, namely Leucaena leucocephala, Gliricidia, Erythrina, and when necessary by oil cakes, e.g. cotton seed cake.

2) Cassava (Manihot esculenta) is another fodder crop of great interest as a feed resource (Devendra, 1977). Its tubers are a valuable energy source which can also provide glucose at the intestine level as its starch can partially escape the rumen fermentation. Its leaves, exploited either as a green fodder (several cuts before harvesting tubers have proved to be still compatible with a satisfactory tuber yield) or, at the time of harvesting, the tubers

are valuable sources of both PDIN and, to a reasonable extent, PDIA.

3) Another interesting plant is the banana, either considered as a whole plant (when blown down by tropical winds and hurricanes) or as a fruit (starch source) when considering the discarded bananas which remain available on the premises of the conditioning exportation units (Le Dividich et al., 1976). The banana as a basis of the diet is more adapted to beef production in view of its high starch and poor N content. As seen above it may however remarkably complement sugarcane. The whole plant can also be envisaged as the basis of the diet for milk producing animals.

AGRO-INDUSTRIAL BY-PRODUCTS

They can be classified into 4 groups:

- By-products providing essentially easily fermentable energy through digestible cell-walls, starch or sugars. They may constitute the basis or the major part of the diet. They derive mainly from sugarcane, citrus, roots and tubers, bananas, coffee...

- By-products which are mainly used as a source of supplementary protein: oil-seed cakes, animal wastes from slaughterhouses and fisheries, by- products from pulses, single cell proteins.

- By-products providing both energy and protein: eg: cereal milling by-products, brewer's and distiller's grains and whey.

- other by-products coming from fruit, bakery and other food industries which provide various kinds of nutrients.

All these by-products have been reviewed by various authors (Chenost and Meyer 1977, IDRC and ICAR 1988). We will therefore restrict to a brief account of the more important ones taken as examples.

Molasses

This is a feed which is rapidly and entirely fermented in the rumen. Between 10 and 30% of the diet, as is traditionally the case there is no particular problems with molasses for all types of livestock. However when the diet is based on molasses (eg: >70%) the behavior of cattle is different and the management of the herd must be more careful (Preston and Willis 1974). A small amount of fibre is vital for ensuring the normal physical function of the rumen. Non-protein nitrogen is essential for the development of the micro-organisms of the rumen. Furthermore the animal responds dramatically to small amount of protein like fish meal, which can escape the rumen fermentation (Preston 1985).

However it has never been possible to incorporate as high levels of molasses in the diet of lactating cows as in the case of fattening cattle. The reason is that diets high in molasses lead to insufficient amount of glucose and glucose precursors (low propionate and high butyrate) in the end products of digestion (Leng and Preston 1976).

Molasses which is an excellent carrier for urea as a source of non protein nitrogen for ruminants can be more easily used as a supplement and distributed to small farmers when is part of solid multinutrient blocks (Leng 1984, Sansoucy 1986, Sansoucy et al 1988).

Citrus and sugarbeet pulps

Due to the high digestibility of their non lignified cell walls they favor, as opposed to molasses, the cellulolytic activity of the rumen. Due to the relatively moderate rate of fermentation (as opposed to sugars) they also represent good carriers of NPN and ensure an efficient microbial synthesis (synchronization of both ATP and NH3 releases). They can constitute the major part of the diet as well as an excellent energy supplement for diets based on fibrous crop residues.

Oil cakes and seeds and by-products of animal origin

They have been comprehensively reviewed in the 1988 IDRC and ICAR's publication. They constitute the largest source of supplementary protein. As mentioned earlier in this paper, the assessment of their potential use as protein supplement will be based on the degree of degradability of their nitrogen in the rumen.

As they represent a source of foreign exchange and of high quality protein for human and non ruminant animals, their use as protein supplement for ruminants should be considered against the local availability of legumes and or legume trees.

Cereal milling by-products

They are very well known and their use is expanding. Their major asset is the fact that they supply at a time, moderately fermentable energy, dietary protein and neoglucogenic nutrients. As an example, rice polishings can play a remarkable role in balancing sugarcane based diets.

CONCLUSIONS

As a main concluding remark and as clearly observed by Preston and Leng (1987), the tropical basic feed resources have in common the fact that they are poor in nitrogen (namely in protected dietary PDIA) and rich in carbohydrates. These carbohydrates are however either structural and slowly fermentable or too easily fermentable compared to those of the temperate fodder plants, rich both in cell wall type and in less fermentable type of energy. As a result, taken alone or in combination with each other, they will be fermented in the rumen at very different rates. In addition there is another drawback in that fermentable N (predominant in the main tropical feedstuffs) may also be released too quickly and not in time with the energy. Supplementing tropical feeds with crop residues, feed crops and agro-industrial by-products, will therefore have to take into consideration not only the above described characteristics but also the kinetics of release of the various nutrients. Attention to the rationing aspects will be of major importance.

REFERENCES

Balch, C.C. 1977. The potential of poor-quality agricultural roughages for animal feeding. In "New Feed Resources". Proceedings of a technical consultation held in Rome, 22-24 November 1976. FAO Animal Production and Health Paper No. 4, pp. 1-15.

Chenost, M. 1975. La valeur alimentaire du Pangola (Digitaria decumbens) et ses facteurs de variation en zone tropicale humide. Annales de Zootechnie 24: 327-349.

Chenost, M. and Mayer, L. 1977. Potential contribution and use of agro-industrial by-products in animal feeding. In "New Feed Resources". Proceedings of a technical consultation held in Rome, 22 - 24 November, 1976. FAO Animal Production and Health Paper No. 4, pp. 87-110.

Devendra, C. 1977. Cassava as a feed source for ruminants. In "Cassava as Feed". Proceedings of a workshop held at the University of Guelph, 18-20 April 1977. pp. 107-119. Ed. Nestle, B. and Graham, M., IDRC.

Evans, T.R. 1977. The potential for increasing cattle production in Malaysia from improved tropical pastures. In Feedingstuffs for Livestock in South-east Asia: pp 37-55. ed. Devendra, C. and Hutagalung, R.I.

FAO. 1985. Better utilization of crop residues and by-products in animal feeding: research guidelines 1. State of knowledge. FAO Animal Production and Health Paper 50.

Göhl, Bo. 1981. Tropical Feeds. FAO Animal Production and Health Series, No. 12, 529 pp.

IDRC and ICAR. Non-conventional feed resources and fibrous
1988 agricultural residues. Strategies for expanded utilization. Proceeding of a consultation held in Hisar, India, 21-29 March 1988. Ed. Devendra, C.

INRA. Alimentation des bovins ovins et caprins. Ed. INRA, 147, rue de
1988 l'Université, 75007 Paris, 471 pp.

Jarrige, R. Place of herbivores in the agricultural ecosystems. In:
1979 Proceedings of the 5th International Symposium on Ruminant Physiology, held at Clermont-Ferrand, on 3-7 September, 1979. Ed. Y. Ruckebush and P. Thivend, pp. 763-823.

Kawamura, O., Tanaka, S. and Miaki, T. The digestibility, intake and
1985 rumen fermentation by sheep of some tropical grasses in Japan. Proceedings of the XV Internat. Grassl. Cong.: Kyoto 24-31 August 1985. pp. 1034-1036.

V.L. Kossila. Global review of the potential use of crop residues as
1985 animal feed. In: Better utilization of crop residues and by-products in animal feeding: research guidelines 1. State of knowledge. pp. 1-13.

Lane, I.R. The use of cultivated pastures for intensive animal
1981 production in developing countries. In A.J. Smith and R.G. Gunn (eds.) Intensive Animal Production in Developing Countries, pp. 105-143. Occasional Publication No. 4. British Society of Animal Production..

Le Dividich, J., Geoffroy, F., Canope, I. and Chenost, M. Using waste
1976 bananas in animal feeding. World Animal Review, 20: 22-30.

Leng R.A., The potential of solidified molasses-based blocks for the
1984 correction of multinutritional deficiencies in buffaloes and other ruminants fed low quality agro-industrial by-products. In Proceedings of the FAO/IAEA meeting on the use of nuclear techniques to improve domestic buffalo production, 30 Jan.-3 Feb. 1984, Manila, the Philippines, pp. 135-150.

Leng R.A. and Preston T.R., Sugarcane for cattle production: present
1976 constraints, perspectives and research priorities. Tropical Animal Production 1: 1-22.

Minson, D.J. Nutritive differences between tropical and temperate
1976 pastures. In: F.H.W. Morley (ed.). World Animal Sciences. Vol. 16. Grazing animals Ch. 7. Amsterdam: Elsevier.

Osbourn, D.F. The utilization of natural grasslands in the tropics.
1976 In: A.J. Smith (ed.). Beef cattle Production in Developing Countries, pp. 184-203. Centre for tropical Veterinary Medicine, University of Edinburgh.

Preston, T.R. and Leng, R.A. Sugarcane as cattle feed. Part 1:
1978 Nutritional constraints and perspectives. World Animal Review 27: 7-12; Part II: Commercial application and economics. World Animal Review 28: 44-48.

Preston, T.R. Nutritional limitations associated with the feeding of
1982 tropical forages. Journal of Animal Science 54: 877-884.

Preston T.R., Validity of feeding standards and development of feeding
1985 systems based on crop residues and agro-industrial by-products. In: Better use of crop residues and agro-industrial by-products in animal feeding: Research guidelines - 1. State of knowledge. Animal Production and Health Paper No 50 pp 197-213. FAO, Rome.

Preston, T.R. and Leng, R.A. Matching ruminant production systems
1987 with available resources in the tropics and sub-tropics. Penambul Books: Armidale Australia. pp 245.

Preston T.R. and Willis M.B. Intensive beef production (2nd edition)
1974 Pergamon Press Oxford.

Ramihone, B. Facteur limitant la fermentation dans le rumen des
1987 pailles de céréales (à petit grain) traitées à l'ammoniac anhydre. Thése, Université des Sciences et Techniques du Languedoc, Montpellier, France, 85 pp.

Salette, J.E. Nitrogen use and intensive management of grasses in the
1970 wet tropics. Proceedings of International Grassland Congress. Surfers' Paradise. pp. 404-407.

Sansoucy R. The Sahel: Manufacture of molasses-urea blocks.
1986 World Animal Review 57:40-48.

Sansoucy R., Aarts G. and Leng R.A. Molasses-urea blocks as a
1988 multinutrient supplement for ruminants. In: Sugarcane as feed. Animal Production and Health Paper No 72 (Sansoucy R., Aarts G. and Preston T.R., editors) pp 263-278. FAO, Rome.

Silva, A.T. and Ørskov, E.R. The effect of five different supplements
1988 on the degradation of straw in sheep given untreated barley straw. Animal Feed Science and Technology, 19: 289-298.

Stobbs, T.H. Factors limiting milk production from grazed tropical
1976 pastures (Limitations to milk production). International Seminar Animal Production, Acapulco, pp. 183-198.

Sundstøl, F. and Owen, E. Straws and other fibrous by-products as
1984 feed. Elsevier, Amsterdam, pp. 604.

Van Es, A.J.H. Nutrition and feeding aspects of using lignocellulosic
1985 materials for animal feed - Proceedings of a OECD Workshop on "Improved utilization of lignocellulosic materials with special reference to animal feed", pp. 54-64, held in Braunschweig 19-21 September 1984.

Van Es, A.J.H. and Taminga, S. Intake and composition of tropical
1987 feeds. In "Milk - The vital force" pp. 573-583. Proceedings of the XXII International Dairy Congress: The Hague 29 September - 30 October 1986. ed. D. Reidel Publishing Company.

Wanapat, W. and Topark-Ngarm, A. Voluntary intake and digestibility
1985 of two selected tropical grasses by swamp buffaloes. Proceedings XV International Grassland Congress. Kyoto. pp. 959-960.

Xande, A. and Garcia-Trujillo, R. Adaptation of the INRA system of
1985 determination and expression of the nutritive value of forages to the tropical conditions in the Caribbean islands. Proceedings of the XV Internat. Grassl. Cong.: Kyoto 24-31 August 1985: pp. 1062-1063.

FEEDING STRATEGIES FOR IMPROVING MILK PRODUCTION OF DAIRY ANIMALS MANAGED BY SMALL-FARMERS IN THE TROPICS

By

R.A. Leng

INTRODUCTION

Milk is an important component of diets for all humans as it is high in essential amino acids that are most likely to be deficient in diets based on vegetable protein. Although milk is a high-cost source of protein and fat relative to vegetable sources, it is readily saleable particularly in the more affluent urban areas of developing countries. Improving milk production is therefore an important tool for improving the quality of life particularly for rural people in developing countries.

Milk production systems in tropical countries are diverse. At the one extreme the systems are similar to those in most industrialised countries and are based on cows of high genetic potential given "high quality feeds" which include fodder crops/silages and grain and protein concentrates. Milk production per cow is extremely high and technological inputs are high. At the other extreme are systems which are used by the vast majority of small farmers in developing countries and are based on low inputs and productivity per cow is relatively low. These small-farmer systems vary from ones in which cows or buffaloes are fed on crop residues, agro-industrial by-products and roadside grass to beef cattle grazing tropical pastures that are milked once a day, with the calf having access to the dam for the other half of each day. In the latter systems, the pastures available to these 'dual purpose' animals are typical of most tropical grasslands and are relatively low in protein and digestibility.

Every conceivable system between these extremes is used in various parts of the world. However, the small farmer on low milk production systems are those with the greatest potential for improvement and are the target of most aide programmes. In the discussion presented here, the strategies for improving milk production from cows/buffaloes fed tropical pastures, crop residues or fibrous agricultural by-products are discussed.

On these feed resources, overall productivity is low, animals reach puberty at a late age (often 4 years) and inter-calving interval is often 18-24 months, resulting in a small number of dairy animals in a national herd being in milk at any one time. A strategy for improving milk production in these systems has therefore two components. The first is to improve reproductive efficiency of the dairy animals and secondly to improve milk yield and persistency.

The greatest scope for improving a country's milk production is through a strategy which targets improvement of reproductive performance. This cannot be achieved however without increasing milk production per animal. Reducing age at first calving from 5 to 3 years and inter-calving interval from 24 months to 12 months by better feeding management will at least double the number of animals being milked at any one time. In addition, because the same feeding strategy that improves reproductive performance also increases milk production, the improved production per animal is also increased.

FEED RESOURCES AVAILABLE TO SMALL DAIRY FARMERS

The small farmers of developing countries have limited resources available for feeding to their ruminant livestock. They do not have the luxury of being able to select the basal diet but use whatever is available at no or low cost. The available resources are essentially low digestibility forages such as tropical pastures (both green and mature), straws and other crop residues and agricultural by-products which are generally low in protein.

The major criterion for improvement in production is to optimise the efficiency of utilisation of the available fodder resource and not to attempt to maximise animal production. There is little point in knowing the "energy" requirements of a cow or buffalo for milk production, whose requirements are to be met from whatever crop residue is available. It is imperative, however, to understand the requirements for supplements that will provide nutrients that will optimise the efficiency of utilisation of that feed resource.

THE BASIC CONCEPTS

When considering how to optimise the utilisation of the available forages for dairy animals, two basic concepts must be applied as follows:

- To make the digestive system of the cow as efficient as possible by ensuring optimum conditions for microbial growth in the rumen.

- To optimise production by balancing nutrients so that these are used as efficiently as possible for milk production without jeopardising the reproductive capacity of the cow.

Any further increases in production may be obtained by the use of supplements of protein, starch and lipids to provide nutrients for milk production above those obtained when the efficiency of utilisation of the basal feed has been optimised. These supplements should be processed and must by-pass the rumen and become available for digestion in the intestine and in this way provide the nutrients in exactly the correct balance for additional milk production.

The two concepts can be implemented by feeding a combination of non-protein nitrogen (NPN), minerals and by-pass protein. The third component is a relatively new concept which suggests that milk production, once the efficiency of utilisation of the basal feed resource has been optimised, depends upon providing nutrients needed for the components of milk, e.g. the quantity and balance of glucose (for milk lactose), protein and fat, in a form that will by-pass the rumen.

AN APPROACH TO IMPROVING NUTRITION OF LACTATING ANIMALS

The approach taken has been one in which urea/molasses blocks (UMB) have been provided to lactating ruminants to allow a slow, continuous intake of nutrients needed to optimise fermentative digestion in the rumen. By-pass protein supplementation is used to optimise the efficiency of use of absorbed nutrients. The development of both these strategies has gone along similar lines with testing under laboratory conditions being followed by testing on well managed farms and eventual trials under village conditions (see Leng and Kunju, 1989).

Over the last year, all Friesians imported into India and placed under the care of NDDB have been fed according to the strategies proposed (see Leng and Kunju, 1989). The 300 day lactational yield has been, on average, 6000 litres.

In addition, several thousand tonnes of a by-pass protein have been fed to cattle and buffalo under village conditions in various climatic zones. On the basis of this research and the experience gained, the largest feed mill in India producing 300-600 tons of feed per day (Amul Feed Mill, Anand) commenced the production of a new pelleted feed supplement containing 30% protein and with approximately 75% of the protein in a form likely to by-pass the rumen.

BACKGROUND - THE USE OF NPN AND BY-PASS PROTEIN IN RUMINANT DIETS

Definition

- By-pass proteins are defined here as those dietary proteins that pass, intact, from the rumen to the lower digestive tract.
- Digestible by-pass protein is that portion of the by-pass protein that is enzymatically hydrolysed in, and absorbed as amino acids from, the small intestine.
- Over-protected protein is that protein of the by-pass protein that is neither fermented in the rumen, nor digested in the small intestine.

- Metabolisable protein is the digestible by-pass protein plus the digestible protein in the microbes that enter the small intestine.

- Fermentable carbohydrates are those parts of the feed carbohydrate that are degraded by microbial action in the rumen to volatile fatty acid (VFA) plus that entering into the microbes that grow with the energy (ATP) released when VFA are produced.

Protein digestion in ruminants

In different production systems, ruminants consume many types of carbohydrates, proteins and other plant and animal constituents. All digestible carbohydrates are fermented to volatile fatty acids (VFA) plus methane and carbon dioxide by microbial action. Proteins are degraded by microbial enzymes in the rumen to give the same three end-products (i.e. VFA, CO_2 and CH_4) plus ammonia (see Figure 1). In all cases a proportion of the substrate metabolised by microbes is used for synthesis of the microbes.

The microbial fermentation of soluble protein in the rumen is an unavoidable consequence of the ruminant mode of digestion. In the absence of other forms of N, it ensures a supply of ammonia nitrogen for micro-organisms from which they synthesize the protein in their cells. Under many circumstances, it is a wasteful process because high quality proteins are broken down to ammonia, absorbed as such, converted to urea in the liver and this is excreted in the urine.

EFFICIENCY OF MICROBIAL GROWTH ON PROTEIN

Protein degradation to VFA leads to a relatively low availability of ATP ('energy') to rumen microbes and therefore protein that is degraded in the rumen is inefficiently used for the growth of micro-organisms. In comparison with carbohydrate when protein is degraded in the rumen, only half the ATP (the energy currency of the microbes) is produced in fermentation of protein relative to the same amount of carbohydrate.

The breakdown of carbohydrate in the presence of adequate ammonia and sulphur and other minerals supplied by, for instance urea/molasses blocks, results in more microbial protein being produced than from an equal amount of protein fermented in the rumen. This is shown diagramatically in Figure 1 and indicates that from a highly soluble protein such as leaf protein, less than 10% of the protein in the diet is available to the animal.

Figure 1. The breakdown of fermentable carbohydrates and protein in the rumen with the production of VFA and microbial protein.

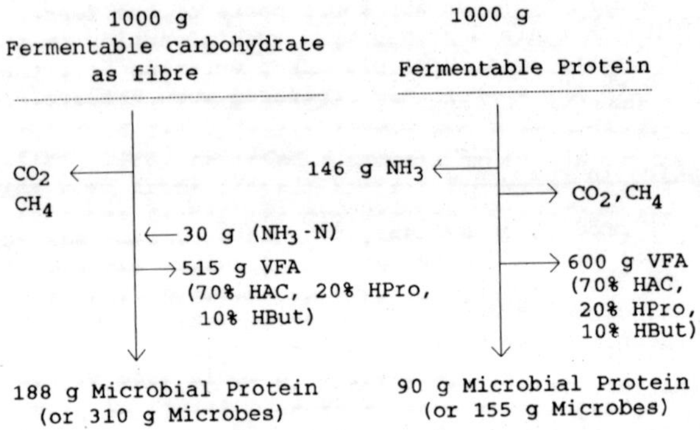

Where HAC = acetic acid
HPro = propionic acid
HBut = butyric acid

Quite clearly therefore with readily soluble and fermentable protein; whilst little escapes the rumen if the protein is in high concentrations the protein to energy ratio in the nutrients arising from the rumen may be decreased.

Factors that influence the availability of by-pass protein

For a variety of reasons a proportion of the dietary protein passes from the rumen into the small intestine without alteration. On reaching the small intestine this by-pass protein is digested by enzyme hydrolysis and absorbed into the body as amino acid.

The conditions under which some dietary protein may escape the rumen for digestion in the lower alimentary tract include:

- When a protein meal has been made highly insoluble by heat treatment.

- The protein meal contains tannins (2-4%) which bind to make an insoluble tannin - protein complex (Barry, 1985) which is not degraded in the rumen but is degraded in the abomasum/small intestine.

- Chemical treatment has been applied, e.g. formaldehyde treatment (Scott, 1970).

- When a relatively soluble protein meal is fed in very high quantities and is either in a finely ground form or is rapidly fragmented into small particles which move quickly through the rumen. For example, when clover or lucerne (that do not contain tannins) are fed at levels below 2.5% of liveweight (on a dry matter basis), it is probable that no dietary protein escapes to the lower tract. However, at levels above this, some protein escapes because of the rapid movement of digesta out of the rumen. The amount of by-pass protein can be as high as 30% of the total protein in the feed if this is highly digestible (D. Dellow & J.V. Nolan - unpublished; Nolan and Leng, 1989.).

- When heat is applied to a mixture of soluble protein and xylose, when a modified browning reaction can insolubilise the protein.

Microbial protein synthesis in the rumen

Ammonia, peptides, amino acids and amines form the nitrogenous substrate for the synthesis of microbial cells but ammonia is the most important source of N for the microbes that ferment forages. Ammonia is used by many species of rumen micro-organisms as their sole source of nitrogen for protein synthesis (see Leng and Nolan, 1984).

This assessment of the role of ammonia in the rumen can be misleading if it is unqualified. Firstly some species of bacteria and protozoa commonly found in the rumen cannot grow or survive unless small quantities of peptides, amino acids or branched chain fatty acids are provided in the diet and are present in low concentration in rumen fluid (Hungate, 1966).

A high level of rumen degradable protein in the diet may support high levels of all N-nutrients needed by bacteria and may cause specific populations of microbes to develop in the rumen as compared to diets where urea alone supplies the fermentable N.

A deficiency of rumen ammonia results in a low microbial growth rate which may reduce digestibility of fibre and lower intake of feed.

The requirements for ammonia for microbial activity

Estimates of the critical level of ammonia in the rumen fluid for efficient digestion has been reported to be as low as 50 mg N/l or as high as 200 mg N/l. However, recent studies have shown that, when ammonia concentrations fall below about 200 mg N/l, the rumen, micro-organisms are inefficient and are likely to respond to dietary NPN supplements particularly to UMB (Krebs and Leng, 1984; Boniface _et al_., 1986; Sudana and Leng, 1986; Perdok and Leng, 1989).

Intake of straw by cattle has been shown to be increased by increasing urea levels in the diet until the level of ammonia reaches 200 mg N/l (Boniface _et al_., 1986; Perdok and Leng, 1989).

Recent studies with buffaloes fed forage based diets showed that, given a period of access to molasses/urea blocks, these animals learn to modify their intake according to the protein content of the basal diet (Table 1).

Table 1. The influence of N content of the basal diet given to lactating buffaloes on the intake of a block lick based on molasses/urea (Leng and Kunju, 1989).

Group No.	Diet N content (gN)	Intake of block lick (g/d)	Milk produced FCM (kg/d)	Liveweight changes (g/d)
1.	0	586	4.3	- 357
2.	30	256	5.7	- 455
3.	83	293	6.3	+ 276
4.	111	173	6.1	+ 89

Can the rumen microbes supply all the protein needs of the ruminant?

Even when ammonia and other nutrients are supplied, the quantities of microbes that leave the rumen in digesta do not supply sufficient protein to meet the needs for productivity in ruminants (i.e. moderate to high growth rates and milk yields). In such a situation, the deficiency symptoms indicate an insufficient supply of essential amino acids to the tissue. Under these conditions supplementation with a protein meal (which has a high content of by-pass protein) to supply additional dietary amino acids increases both the level and efficiency of animal production (see Preston and Leng, 1987).

Protein (or Amino Acid) Requirements of Ruminants

In the past, the protein requirements of ruminants and evaluation of the protein value of feeds for ruminants have been based on digestible crude protein (N x 6.25). This is now recognised as a misleading concept. The use of digestible crude protein has arisen largely because it was considered that cattle and sheep could obtain their essential amino acids from microbes produced in the rumen. This in turn led to suggestions that extensive use could be made of non-protein nitrogen in high carbohydrate feeds and that a special role of ruminants could be to convert non-protein nitrogen to high quality animal protein.

These have now been superseded by new concepts which take into consideration that when amino acid requirements are high, insufficient digestible microbial protein is available from the rumen to meet these needs. It is now necessary to assess the requirements for N by ruminants in terms of the amount of ammonia (or NPN) and amino acids needed by the rumen microbes, and the amount of digestible by-pass protein needed by the animal to augment the total protein (amino acids) available to the animal and to create an efficient metabolism. The sum of the two sources of digestible protein represents the metabolisable protein.

Protein or amino acid requirements relative to energy requirements of ruminants are, however, influenced by a number of factors and cannot be stated with any degree of accuracy. The requirements are influenced by:

- physiological state of the animal,
- rate of growth and milk production,
- body composition as influenced by previous dietary and health history,
- basal feed (particularly fat content),
- proportions of the different amino acids absorbed,
- patterns of rumen fermentation (i.e. acetate:propionate ratio),
- availability of volatile fatty acids,
- requirements for glucose for essential purposes,
- environmental heat or cold stress, and
- the extent of the work load of the animal.

With all these unknowns, the need for by-pass protein under conditions pertaining to small-holder cattle can only be assessed in feeding trials aimed at developing response relationships. The effects of physiological state of the female goat on the utilisation of protein and, therefore its requirements, is well illustrated by the data shown in Figure 2.

Metabolisable Protein Available to Ruminants

Metabolisable protein available is the sum of digestible dietary by-pass protein plus digestible protein from microbes reaching the lower tract. On most straw based diets the metabolisable protein is mainly of microbial origin (i.e. there is no by-pass protein in the diet). The amount of protein available therefore depends on the efficiency of microbial growth in the rumen.

This in turn depends on several factors:

- the presence of all the essential nutrients in the balances and amounts needed by the rumen microbes to grow e.g. ammonia, sulphur, phosphorus, trace minerals, amino acids, peptides, etc.,

- a source of fermentable dry matter, i.e. the feed consumed,

- to a small extent the rate of digesta turnover and therefore feed intake. However, this depends on degradability of the feed, type of carbohydrate and the physiological status of the animal.

- buffering capacity of the rumen and pH of the rumen fluid which largely depends on diet, and

- the balance of micro-organisms in the rumen. If supplementation with carbohydrate promotes protozoal population this can actually decrease the protein to energy ratio in the nutrients available from the rumen (see Bird and Leng, 1985).

As an example of how the balance of microbial protein to VFA energy can be altered in a cow given a straw based diet, the effects of an inefficient rumen (low rumen ammonia supply) and an efficient rumen (optimum rumen ammonia) are shown in Table 2. (see Leng, 1982 for the assumptions and calculations).

The point is that the P:E ratio in the nutrients absorbed is altered according to how efficiently the rumen organisms are digesting the feed or how much by-pass protein there is in the diet.

Figure 2. The effects of physiological state on the intake and retention of nitrogen in goats fed oaten hay/lupins (11% crude protein) (Halais, 1984)

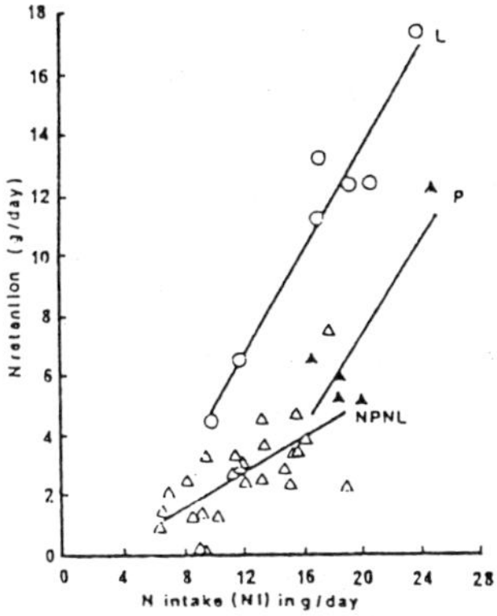

Table 2 The theoretical effect of feeding urea and urea plus by-pass protein on the P:E ratio in cattle. The values were calculated for a bovine consuming 4 kg of digestible organic matter without or with urea or with urea and 400 g of a by-pass protein source.

Rumen condition	Microbial protein synthesised (g/d)	Total protein available (g/d)	VFA produced (MJ/d)	*P/E ratio (g protein /MJ VFA)
Deficient in ammonia	500	500	41	12:1
Sufficient in ammonia	1010	1010	30	34:1
Ammonia sufficient + 10% of the diet as by-pass protein	1010	1410	30	47:1

* no consideration is taken here of the digestibility of the microbial or dietary by-pass protein

Effect of increasing ammonia concentrations in the rumen of cattle on N deficient diets

In most situations, adding urea to a low protein diet, such as that based on a cereal crop residue, increases intake of the basal diet in addition to improving microbial growth and digestibility (Table 3).

Table 3. The effect of infusing urea into the rumen of a cow given straw based diets (Campling *et al.*, 1962).

Diet	Straw DM Digestibility (%)	Intake of Straw (kg/d)	Theoretical* P:E ratio (g protein/MJ VFA
Straw	39	5.6	12:1
Straw + 150 g urea	47	7.9	34:1

* taken from Table 1.

The potential effects of providing a UMB to ruminants on low protein forages (which is intended to provide urea and other nutrients) include the following:

- Increased digestibility of straw
- Increased feed intake
- Increased absorption of total nutrients
- Increased P:E ratio in the nutrients absorbed

The effects of supplementation of by-pass protein

Supplementing a diet of crop residues fed to cattle with a by pass protein improves the P:E ratio in the nutrients absorbed (see Table 1 and 2). This has a large influence not only on the level of production but on the efficiency of feed utilisation (i.e. the amount of feed required per unit of milk production or growth, is lowered). Stated in another way, animals produce less metabolic heat when P:E ratios are well balanced to requirements. This is well illustrated by research shown in Table 4 where straw intake has been maintained constant and efficiency of utilisation of the feed is improved by supplementation. In other studies the increased efficiency is not readily discernible as the effect of such supplements is to increase forage intake (see Preston and Leng, 1987).

Table 4. The growth rate of calves (live weight 150 kg) given rice straw and supplemented with an oilseed meal (Saadullah, 1984).

Daily Supplement (g/d)	Straw intake (kg/d)	Liveweight-gain (g/d)	Feed conversion ratio (kg feed/kg gain)
0	3.8	84	46:1
200	3.8	371	11:1
400	3.8	373	12:1
600	3.8	508	9:1

RESEARCH ILLUSTRATING THE RESPONSES OF CATTLE TO UREA/MOLASSES BLOCKS AND BY-PASS PROTEIN MEAL SUPPLEMENTATION

Growth studies

Jersey bulls (350 kg live weight) fed rice straw plus a concentrate (low in true protein i.e. about 15%) trebled their rate of weight gain when fed a molasses/urea block in conjunction with 1 kg of this concentrate (Table 5).

Studies with lactating cows/buffaloes

In ten villages, the average milk sold in the collection centres increased by 0.4-1.1 litres/day when the farmer made a molasses/urea block available to their dairy buffaloes (Table 6). Other trials showed that concentrate supplementation could be reduced without loss of milk production when a molasses/urea block was given.

Table 5. The effects of supplying molasses/urea blocks to cattle fed rice straw plus 1 kg 15% concentrate (Kunju, 1986).

	Straw intake (kg/d)	Block intake (g/d)	Live Wt. change (g/d)	Feed cost/kg gain (Rupee/kg)
No block	6.4	0	220	9.3
With block	6.8	530	700	3.7

Table 6. The observations on response of feeding block licks in villages (Kunju, 1986). The results show the milk or milk fat sold to the local collection centre (Kaira District Co-operative Milk Producers' Union Ltd., Anand, India).

Village	Milk (kg/d)		Milk fat (g/d)	
	pre lick	with lick	pre lick	with lick
Alwa	4.8	5.9	300	450
Punadhara	4.0	4.8	270	340
Fulgenamuwada	2.4	3.5	160	280
Hirapura	4.2	5.2	350	480
Bamroli (N)	3.6	4.2	270	380
Dehgam	4.3	4.7	310	350

More recently it has been demonstrated that feeding a meal high in by-pass protein (low in grain) as compared to a cattle feed concentrate based on traditional requirements increased milk production and live-weight gain without substantially influencing basal feed intake (Table 7).

The cattle were each fed 40 kg of green forage daily. The forage consisted of 60% legume (mostly lucerne/cowpea) and 40% non-legume (maize, sorghum/oats). The concentrates for cattle in group 1 were fed according to NRC recommendations. The cattle in group 2 were fed a protein concentrate based largely on solvent extracted protein meals demonstrated to have a high by-pass protein content. A major point here is that the animals in group I disposed of nutrients equivalent to 20-25 MJ of energy presumably through 'futile cycles of metabolism'. This additional metabolic heat production could have increased body temperature by 16.5°C if the animal had been in an environment where this extra heat could not have been dissipated. The feeding trial was conducted during the cool season but

Table 7: The effects of replacing balanced concentrates with a high by-pass protein pellet on live-weight change and milk yield of Jersey x Kankrej cows (M.G.P. Kurup, G. Kunju NDDB, India - pers. comm.).

Group	No./group	Crude Protein in supplement (%)	Intake of supplement (kg/d)	Milk Yield (kg/d)	Live-weight change (g/d)
1	75	18	4.7	8.0	- 210
2	75	30	2.6	8.8	+ 202

clearly in the hot season feed intake could not have been maintained. Put another way, if the environmental temperature was critical for cattle in group II then the animals in group I would have needed to reduce their feed intake by 20 MJ ME.

CONCLUSION DRAWN FROM STUDIES IN INDIA

The efficiency of feed utilisation is enormously improved if the rumen of the animal has a healthy microbial population adequately supplemented by providing a molasses/urea block which often increases the intake of a basal diet. Adding a by-pass protein supplement will further improve the efficiency of utilisation of the basal feed resources but will also allow animals to maintain feed intake at high environmental temperatures and humidity. Conversely, the productivity of lactating animals can be maintained at a lower feed intake provided the rumen is made efficient and the animal's metabolism is made efficient by supplementing with a molasses/urea blocks and by-pass protein meal respectively.

CONSTRAINTS TO APPLICATION OF THE BY-PASS PROTEIN TECHNOLOGY

Even though the application of UMB/by-pass technology is highly promising, a few constraints are still to be overcome before it can be widely applied with confidence. Some of these are given below and indicate areas for intensive research:

1. The information regarding the degradabilities of protein in all raw materials used in cattle feed are not yet available and may be quite variable depending on source, manufacturing conditions and presence of other compounds.

2. Easy laboratory tests for protein degradability are still not available and there is still some considerable disagreement as to which method provides the best indication of the content of by-pass protein in a protein meal.

3. There are insufficient data from feeding trials available on milk production per unit input of by-pass protein under the systems commonly used by small farmers.

4. There are no response relationships for milk production for economic analysis of the feeding of by-pass proteins which covers at least two lactations. This is important as by-pass protein supplementation on these diets often improves the body condition of cattle and therefore reproductive performance. The second lactation after introduction of these systems may show the greatest economic response.

5. Many protein meals are undegradable in the rumen. However, their digestibility in the intestines may be very low. This applies particularly to protein meals with high tannin content. Such protein meals are not good sources of protein to the animal since much of the protein is lost in faeces.

6. For the most efficient utilisation of by-pass protein for production, the essential amino acid to total N ratio must be high.

7. The limits of responses to by-pass protein resides in the digestible energy content of the diet and at low digestibilities, high level feeding of a by-pass protein meal will result in amino acid degradation as an energy supply.

PRACTICAL APPLICATION OF BY-PASS PROTEIN IN VILLAGE SOCIETIES

Feeding Friesian cows of high genetic potential for milk production - The National Dairy Development Board of India (NDDB) experience

Friesian cows of German origin were imported into India as potential mothers for the next generation of bulls for cross-breeding with indigenous cows. These animals were distributed to (1) NDDB farms with management and accurate recording of milk yield and (2) individual village farmers in cool environments. The NDDB farms, which are situated at Anand and Bidaj in Gujarat, are in areas with extremely high summer temperatures which often exceed 40°C and may at times exceed 50°C.

All animals are fed whatever forage is available and were provided with urea/molasses blocks and fed only a by-pass protein concentrate (30% CP) at 300-500 g/litre of milk production. All animals have thrived, most are now in their second lactation and where accurate records have been kept have produced between 6000- 6900 litres of milk per 300 day lactation with peak daily lactations often exceeding 30 l/day.

The point that has to be emphasised is that these animals were apparently relatively unaffected by the hottest period of the year and maintained milk production at a time when there is usually a marked reduction in milk yield. They were fed the available forage which varied from mixtures of rice straw and green oats/crops through to a mixture of rice straw and tropical grass. The practical observations support the more controlled research under institutional/ laboratory conditions and indicate a major influence of balancing nutrition on amelioration of heat stress in lactating animals.

Amelioration of Anoestrous in Village Buffalo/Cattle

A major problem associated with milk production in village societies is that the "non-descript" animals which are by far the majority of dairy animals are often fed the poorest feeds particularly in early life and between lactations. The reason for this is that without the cash flow that comes from milk and with no rapid cash return on their outlay, village people (who always experience cash flow problems) are not prepared to purchase supplements.

In general, in developing countries, cattle and buffalo often calve for the first time at 4-5 years of age and have an inter-calving interval of up to two years. Infertility is therefore a major problem.

The improvements in growth rates mediated by the feeding strategies discussed here also suggest that reproductive rate may be similarly improved. A demonstration trial was established to test this hypothesis. Within two village societies, cattle and buffalo were selected that had exhibited (over an 8-12 month period) either infantile genitalia (buffalo heifers) or no ovarian activity in mature cows/buffaloes. These animals were provided with molasses/urea multi-nutrient blocks over the hot summer months and 90% of these animals came into oestrous after 3-4 months (Table 8). These studies have also been supported by studies of grazing cows supplemented with molasses/urea blocks in Africa which have shown a marked decrease in the lactational anoestrous period (Table 9).

The implications for improving milk production of these discoveries is extremely large. Decreased age at first calving, together with decreased inter-calving interval, may increase the total number of animals lactating at any one time by 2 or even 3 fold, this in turn will increase milk production from the national herd by the same increase.

Table 8. The effects of providing molasses/urea blocks to cattle and buffalo on reproductive activity (John, NDDB, personal communication). The animals were owned by small-farmers and had been diagnosed as anoestrous (adult animals) or having infantile genitalia (buffalo heifers) and had been in this condition for 8-12 months. The farmers were given molasses/urea multinutrient blocks at no cost. The period covered was the hottest part of the year.

	No.	No. of animals coming into oestrus		
		before 120 d	after 120 d	never
Crossbred cow	12	11	1	0
Adult buffalo	18	17	1	0
Buffalo heifers	39	28	6	5

Table 9. The effect of providing a molasses urea block (UMB) to grazing cows (Gobe Ranch, Ethiopia) on the length of the post- partum or lactational anoestrous period (ILCA ,1987).

	Post-partum anoestrous period (days)		
	UMB supplement	no supplement	diff.
Continuous suckling	132	199	67
Restricted suckling	113	159	46

TREATMENT OF CROP RESIDUES TO IMPROVE DIGESTIBILITY

The treatment of crop residues with alkalis to improve digestibility is a well researched and established technique. Feeding treated straw as compared to untreated straw considerably improves ruminants productivity (see Sundstøl and Owen, 1984).

Simple techniques based on ensiling the wet straw (50% moisture) with 3-4% urea are well established and could be applied under village conditions. However, these techniques are only being accepted slowly or are unacceptable to small farmers for a variety of reasons which vary from country to country and within districts in the same country. The main constraints to implementing straw treatment as a means of improving milk production in small-farmer systems are economic, sociological and logistic.

Economic considerations

Small farmers invariably have a cash flow problem and purchase of urea is restricted generally for crop production. Often plastic covers for the straw are costly and impractical. In addition, the returns for use of urea on a rice crop must be offset against the income from milk.

Sociological considerations

Often the most appropriate time for treatment of crop residues is at harvest time, when most the family are involved in long hours of work and have no time to treat straw. The availability of water is often a constraint. In most countries this would be carried in urns by the women from a distant source. These wives/daughters of small farmers generally have very full working days. Often, for security or convenience purposes, straw is stored in or close to the residence of the family and the smell of ammonia is highly unacceptable and may lead to eye disorders particularly in children. Finally, wet straw is much more difficult to store, preserve and feed to the cattle.

A major constraint is that farmers, from experience, have a fairly accurate annual feed budget. The main benefit from treated straw comes from increased feed intake and therefore the budget has to be adjusted. Failure to do this often results in the farmer having to purchase expensive straw which will be economically disadvantageous.

Conclusion

For straw treatment to be successfully accepted by small-farmers in developing countries the methods must be made easy, low cost and must have low labour inputs. It seems that, for the foreseeable future, straw treatment is unlikely to develop as a national strategy but will be used by the larger farmers particularly those that can afford to buy labour.

RESEARCH NEEDS

For the implementation of the new feeding system, more feeding trials need to be carried out in which response relationships of milk production/weight change can be correlated with level of by-pass protein feeding. However, some of this can be left to individual farmers who can be instructed to slowly increase the level of protein meal until they are satisfied with the response. They will automatically take the most economic option and the important point to stress is that farmers must have access to the supplements.

The influence of these feeding strategies on reproductive performance needs further research as it is likely to have the greatest effect within a country.

The feed processing technology should be modified in view of the new system with a view to increasing, in processing, the by-pass protein content of a pelleted feed. A suitable feed formula based on the nutrient supply, processability and economics of feeding needs to be developed, for use with the important basal feeds available to small-farmers.

There is a wide gap today in this technology between the research nutritionists who use only single ingredients or a combination of two or three protein meals and the practical feed manufacture who uses a variety of feeds compounded on least-cost basis. Since many developing countries have large quantities of protein meals in the country then technology development to ensure its efficient utilisation should be a matter of priority. In countries where the oilseed meals are unavailable, the potential of forage trees containing tannins, or the treatment of forage tree leaves to protect the protein need to be developed.

THE FUTURE

The challenge for the scientist in many developing countries is to how best combine in a diet for dairy animals the available green forage, crop residues and agro-industrial by-products with the available protein resources and molasses/urea block to optimise milk production. It is likely that the availability of protein for dairy animals is likely to be the primary economic constraint, it is therefore necessary to develop new protein resources (e.g. aquatic plants, tree crops) and to find ways and means of protecting the protein from degradation in the rumen whilst remaining of high digestibility is an urgent priority.

Increasing milk production following optimisation of the efficiency of utilisation of the basal feed resource

The "Greenhouse effect", that is the warming of the Earth's atmosphere because of increased content of carbon dioxide and methane, will in the future require a reduction in production of these gases. Methane produced by ruminants probably contributes about 25% of the increase in global methane concentration in the atmosphere (which is 1% per year at present) and this source of methane can be reduced by decreasing the number of ruminants in the world. This will necessitate a move to increase production per animal to maintain and increase this source of human food. This increase per animal will need to be made within the constraints of the available feed resources.

Milk is essentially water, lactose, protein and fat. To boost production of milk from animals fed available forages above that stimulated by the optimum level of by-pass protein plus urea/molasses blocks, it will be necessary to supplement well balanced mixtures of amino acids (from by-pass protein) and lipids (as unreactive LCFA combined with calcium to form soaps) and by-pass starch.

The role of dietary fat in the nutrition of ruminants has traditionally been looked upon as a means of increasing the energy intake of ruminants without a proportional increase in the quantity of feed consumed. A strong case for inclusion of fats in ruminant diets was made by Milligan (1971) on the basis of the energetic efficiency of incorporation of dietary long chain fatty acids (LCFA) into tissue LCFA of fattening animals. Kronfeld (1976,1982) proposed that an optimal balance between aminogenic, glucogenic as well as lipogenic nutrients is required for maximal efficiency of milk production and prevention of ketosis in highly productive dairy cows. Theoretically, this should be achieved when, amongst others, exogenous LCFA contribute 16% of the total ME intake (Kronfeld, 1976). Similar levels of LCFA inclusion in the diet of lactating cows, was determined to result in an optimal efficiency of nutrient utilization for milk production by Brumby *et al*. (1978).

Very little information is available, however, on the influence of dietary LCFA on the efficiency of nutrient utilization by growing and lactating ruminants, especially when they are fed roughage-based diets. Although results reported in the literature are highly variable, generally it is believed that the inclusion of more than 4-6% fat in the diet will result in a reduced digestibility of fibre in the rumen and sometimes a reduced DMI (Kronfeld, 1982; Moore *et al*., 1986), unless these lipids are offered in a form which makes them relatively inert in the rumen.

Calcium salts of LCFA (Ca-LCFA) are such a source of ruminally inert LCFA (Palmquist and Jenkins, 1982; Jenkins and Palmquist, 1984)

and have been shown to increase milk production by dairy cows when used as a feed supplement (Palmquist, 1984). Interactive effects of dietary LCFA with nutrients other than fibre have received little attention.

An interaction between dietary LCFA and protein meals has been found and on low protein diets, the benefits of dietary fat are only apparent when a by-pass protein is fed (van Houtert and Leng, 1986).

The quantitative importance of these possible nutrient interactions is unknown in dairy animals fed forage based diets but since the nutrients in milk arise from long chain fatty acids, amino acids and glucose, research is now being aimed at developing a supplement which provides directly to the animal.

REFERENCES

Barry, T.N. 1985. The role of condensed tannins in the nutritional value of *Lotus pedunculatus* for sheep. 3. Rates of body and wool growth. *Brit. J. Nutr.* 54: 211-17.

Bird, S.H. and Leng, R.A. 1985. Productivity responses to eliminating protozoa from the rumen of sheep. In: Biotechnology and Recombinant DNA Technology in the Animal Production Industries - Reviews in Rural Science 6. Eds. R.A. Leng, J.S.F. Barker, D.B. Adams and K.J. Hutchinson University of New England Publishing Unit, Armidale, Australia.

Boniface, A.M., Murray, R.M., Hogan, J.P. 1986. Optimum level of ammonia in the rumen liquor of cattle fed tropical pasture hay. *Proc. Aust. Soc. Anim. Prod.* 16: 151-54.

Brumby, P.E., Storry, J.E., Bines, J.A. and Fulford, R.J. 1978. Utilisation of energy for maintenance and production in dairy cows given protected tallow during early lactation. *J. Agric. Sci., Camb.* 91: 151.

Campling, R.C., Freer, M. and Balch, C.C. 1962. Factors affecting the voluntary intake of food by cows. 3. The effect of urea on the voluntary intake of straw. *Br. J. Nutr.* 16: 115.

Halais, C. 1984. Nitrogen retention in female goats. *Proc. Nutr. Soc. Aust.* 9: 149.

Hungate, R.E. 1966. The rumen and its microbes. Academic Press, New York.

ILCA. 1987. International Livestock Centre for Africa. Annual Report. p.4.

Jenkins, T.C. and Palmquist, D.L. Effect of fatty acids or calcium
1984 soaps on rumen and total nutrient digestibility of dairy rations. J. Dairy Sci. 67: 978.

Krebs, G. and Leng, R.A. The effect of supplementation with
1984 molasses/urea blocks on ruminal digestion. Proc. Aust. Soc. Anim. Prod. 15: 704.

Kronfeld, D.S. The potential importance of the proportions of
1976 glucogenic, lipogenic and aminogenic nutrients in regard to the health and productivity of dairy cows. Advances in Animal Physiology and Animal Nutrition 7: 5.

Kronfeld, D.S. Major metabolic determinants of milk volume, mammary
1982 efficiency and sponteneous ketosis in dairy cows. J. Dairy Sci. 65: 2204.

Kunju, P.J.G. Urea-molasses block lick: a feed supplement for
1986 ruminants. In: Rice Straw and Related Feeds in Ruminant Rations M.N.M Ibrahim and J.B. Schiere Proc. of Workshop, Kandy, Sri Lanka pp. 261-274.

Leng, R.A. Modification of rumen fermentation. In: Nutritional
1982 Limits to Animal Production from Pastures. J.B. Hacker, ed.) CAB, Farnham Royal, U.K. pp. 427.

Leng, R.A. and Kunju, G. Development of feeding strategies for
1989 improving milk production in India from milch animals owned by small-farmers. In: Final Co- ordinated Meeting of "The Use of Nuclear Techniques to Improve Domestic Buffalo Production in Asia" IAEA - Rockhampton, Australia.

Leng, R.A. and Nolan, J.V. Nitrogen metabolism in the rumen. J.
1984 Dairy Sci. 675: 1072.

Moore, J.A., Swingle, R.S. and Hale, H.W. Effects of whole
1986 cottonseed, cottonseed oil or animal fat on digestibility of wheat straw by steers. J. Anim. Sci. 63: 1267.

Milligan, L.E. Energetic efficiency and metabolism transformations.
1971 Federation Proceedings 30: 1454.

Nolan, J.V. and Leng, R.A. Manipulation of the rumen to increase
1989 ruminant production. Proceedings of Combined Consultants and Research Co-ordination Meeting, 13-17 March, 1989, I.A.E.A., Vienna. pp. 149-166.

Palmquist, D.L. and Jenkins, T.C. Calcium soaps as a fat supplement
1982 in dairy cattle feeding. In: Proceedings of the 12th World Congress on Diseases of Cattle, Amsterdam.

Palmquist, D.L. Use of fats in diets for lactating dairy cows. In:
1984 Fats in Animal Nutrition. J. Wiseman, ed. Butterworths, London, pp. 357.

Perdok, H. and Leng, R.A. Rumen ammonia requirements for efficient
1989 digestion and intake of straw by cattle. In: The Roles of Protozoa and Fungi in Ruminant Digestion. Eds. J.V. Nolan, R.A. Leng, D.I. Demeyer. Penambul Books, Armidale, Australia. In press.

Preston, T.R. and Leng, R.A. Matching Ruminant Production Systems
1987 with Available Resources in the Tropics and Subtropics. Penambul Books, Armidale.

Saadullah, M. Studies on utilisation of rice straw by cattle. Ph.D.
1984 Thesis, Royal Veterinary University, Copenhagen.

Scott, T.W. Protected lipid-protein supplements. In: Feeding
1970 Protected Proteins to Sheep and Cattle. Proc. of the Australian Society of Animal Production Meeting, September 1970. pp. 35-53.

Sudana, I.B. and Leng, R.A. Effects of supplementing a wheat straw
1986 diet with urea or a urea-molasses block and/or cottonseed meal on intake and liveweight change of lambs. <u>Anim. Feed Sci. Technol.</u> 16: 25.

Sundstøl, F. and Owen, E.C. In: Straw and other fibrous by-products
1984 as feed. Elsevier, Amsterdam.

van Houtert, M.F.J. and Leng, R.A. Strategic supplements to increase
1986 the efficiency of utilisation of rice straw by ruminants. In: Rice straw and related feeds in ruminant rations. Eds. M.N.M. Ibrahim and J.B. Schieve. pp. 282- 284. Agricultural University, Wageningen.

FEEDING RIVERINE BUFFALOES FOR MILK/DUAL PURPOSE PRODUCTION

by

A.M. EL-Serafy

INTRODUCTION

Of the world total of about 138 million buffaloes (Jasiorowski, 1988), riverine buffaloes represent 70% with the concentration in India (76 M) Pakistan (14 M) and Egypt (2.4 M). Jasiorowski (1988) and Mudgal (1988) point out an increase of about 11.2% in the number of the river buffalo type between 1983 and 1986, indicating an increasing awareness by farmers of the importance of the animal in their economic life and as an integral part of the farming system.

Proportions of buffalo milk and meat respectively in the world total production increased from 5.5% and 0.8% in 1976 to 6.8% and 1.0% in 1986. In Asia and the Pacific countries, total milk production from buffaloes in 1985 was about 31 million tonnes (91% of total) (Mudgal, 1988) with India alone producing about 22 million tonnes. Corresponding values for meat production are 1 million (80%) and 0.3 million tonnes in Pakistan. In Egypt, buffaloes produce 65-70% and 45-50% of total milk and meat respectively, (Central Agency for Statistics, 1986).

For many decades however, research for development was slow, scattered and uncoordinated and failed to achieved meaningful results. Consequently, it was believed that the low fertility and low production levels are inherent traits of the species. Fortunately, coordinated research in the past 25 years in Egypt, India, Japan and Taiwan supported the theory that low production levels are related to poor management and to poor nutrition in particular.

The main aims of this paper are :

a) to compare the characteristics of the digestive physiology and nutrition of the buffalo and the cow, and
b) tc describe an improved feeding/management package for enhancing the production of milk and meat.

COMPARATIVE DIGESTION AND NUTRITION OF RIVERINE BUFFALOES AND CATTLE

There is no difference in the digestive tract between the buffalo and the cow, the four-pouched stomach and the rest of the gastro-intestinal tract being the same in both species. The rumen in cows and buffaloes is well adapted to utilize the cellulosic matter and the main fermentative compartment proceeds the main site of digestion, allowing the maximal use of fermentation products.

From a functional point of view however, there might be a difference between the riverine buffalo and the cow in ability to digest poor quality roughage, e.g. rice straw (Ranjhan, 1988). The reason for this difference, reported from feeding trials, is not quite understood although differences in rumen bacterial growth rate between species were reported by Zaki El-Din et al. (1985) as a result of feeding the same roughage diet with or without added urea and/or molasses. The ability of the buffalo to consume more DM from rice straw than the cow could further explain the difference in digestion (Devendra 1987).

In Egypt, research on comparative digestibility and efficiency of feed utilisation between buffaloes and cows is limited. It has been reported that local buffaloes and cows digest concentrates and good quality roughages, like berseem hay, equally well. With poor quality roughages like rice straw however, the buffalo excelled the cow in digesting DM and CF (El-Ashry 1988; Saied Mahmoud, personal communication). With regard to the efficiency of feed utilisation for meat production, the buffalo steer calves produced more meat per unit of feed intake than either local steers of native or Friesian breeds (El-Ashry et al., 1975).

Research reports from India indicated the superiority of riverine buffaloes over cows in lignin turnover and that was due to animal size being responsible for greater digestion in buffaloes than cows (Mudgal, 1988). The results also show that TDN output/input ratio varied from 6 to 30% and protein output/input ration from 5 to 40%, indicating that buffaloes fed on straw and a grain-based diet were more efficient than cows. With regard to comparative utilization of energy for milk production, it has clearly been shown that maintenance and production requirements were higher in Murrah buffaloes than in Brown Swiss x Sahiwal cows, indicating that cows were more efficient in utilizing metabolizable energy for milk production than buffaloes (Mudgal, 1988).

FEEDING/MANAGEMENT OF RIVERINE BUFFALOES

Feeding from birth to weaning

A project was started in 1973 and continues at Ain Shams University, Faculty of Agriculture to study the effect of improved feeding management on the performance of buffalo calves during pre- and post-weaning phases of growth. The accumulated results from this project (El-Bassioni, 1983; El-Serafy et al., 1982) and from other research stations in Egypt were summarized by El-Serafy and El-Ashry (1989).

In general it was concluded that, to achieve maximum benefits from rearing calves on milk replacers, a package of management is required, namely to feed restricted amounts of replacers (4 kg liquid divided into two meals), to have fresh water available, to avoid using antibiotics in milk replacers, to introduce a mash starter from two weeks, to offer good quality berseem hay leaves ad libitum and to rear calves in a well ventilated barn, always using a dry bed of rice straw.

Feeding buffalo males for growth

From weaning to about 150 kg body weight, male calves require special attention in formulating rations to promote maximum tissue growth. A highly digestible pelleted starter (70 to 75% TDN and 15 to 17% DP) is essentially required to achieve about 0.7-0.8 kg ADG (EL-Ashry et al,. 1981). The ratio concentrate to roughage ranges between 50:60 or 60:40 on a DM basis, with good quality berseem hay making up at least half of the roughage (El-Koussy 1981). Comparable ADG values for buffalo calves at the same age/weight reported in the fifties and sixties were much lower (400-600 g: Ragab et al., 1966).

Different roughages fed to male calves during growth have shown the superior effect of rice straw, compared to wheat or bean straws (Afifi, 1977). Their results showed that calves required 4.42, 4.68 and 4.80 kg feed DM to produce one kilogram gain.

Rice straw contains more ligno-cellulose bonds and ash than wheat or bean straws (Van Soest, 1987, personal communication) and its TDN value is less than the other two straws (Abou Raya, 1967). A possible explanation for better efficiency of utilization by buffalo calves is that the rumen cellulolytic micro-organisms in buffaloes are more capable of breaking these bonds, making hydrolyzed glucose units available for VFA production (Abou Akkada and El-Shazly, 1966; Zaki Eldin et al., 1985)

Feeding for fattening of buffalo males

Two fattening practices of male buffalo calves are recognize in Egypt:

a) Fattening from 200 to 350 kg, over a short fattening period of about 4 months, and
b) Fattening from 250 to about 500 kg over a relatively longer period of 10 to 11 months (called bitello).

The first practice produces relatively juicier meat but the second is the main practice because of its high dressing yields.

The overall ADG during fattening is usually between 800 to 900 g/d, depending on the level of concentrates, being higher with concentrates level in ration over 50% of the diet in which the main roughage as rice straw (Afifi, 1977). In feedlot fattening operations (Shehata et al., 1973), ADG was 800 to 1000 g when the concentrate portion of the ration was 75% and when 1 kg concentrates was offered for each 50 kg live body weight.

Carcass measurements

Although in fattening trials different ratios of concentrates to roughages were used, the dressing percentage ranged from 50 to 60%, depending on the weight at slaughter, being higher than 52% at slaughter weights above 400 kg. Also, high dressing percentages, meat:bone ratios and carcass-fat are usually associated with high levels of concentrates (Table 1).

Table (1) Dressing and bone-less meat percentages of buffalo calves slaughtered at different weight categories.

Item	Feeding groups*					
	I	II	III	IV	V	Overall average
Calves slaughtered at 300 kg						
Dressing %	53.3	53.5	57.4	50.2	47.1	52.3
Boneless meat %	83.2	81.7	82.8	79.9	81.7	81.8
Calves slaughtered at 400 kg						
Dressing %	57.1	54.7	54.1	54.0	57.6	55.5
Boneless meat %	81.2	82.5	82.7	81.9	83.4	82.3
Calves slaughtered at 500 kg						
Dressing %	59.8	60.2	59.7	60.6	54.4	59.9
Boneless meat %	79.3	82.1	84.0	83.0	78.2	81.3

* Feeding groups I to V corresponds to levels of Napier grass of 5, 10, 15, 20 and 25%, respectively on DM basis.

In conclusion, the recommended feeding regime for growing male buffalo calves from 90 to 200 kg live body weight consists of a 50:50 concentrate to roughage ratio on a DM basis. The concentrate portion should be highly digestible pellets containing 65-70% TDN and at least 15% DP, while the roughage portion is made up of 2-cut berseem hay and rice straw (50:50 ratio). Intake in this growing period was calculated as 3% of body weight. For fattening purposes, the rations should contain between 65-80% concentrates.

Feeding of growing/pregnant heifers

Raising of a good buffalo heifer is a prerequisite for achieving a high-yielding buffalo cow. The characteristics of a good heifer in Egypt (El-Ashry, 1988: El-Fouly and Afifi, 1977) are to weigh 350-370 kg at 16-18 months of age at which the heifer reaches sexual maturity, exhibit regular oestrus cycles and to be ready for mating in order to deliver her first calf at 27-28 month of age (460 to 480 kg weight).

From weaning to about 180 kg, heifers require the same special attention as was described in feeding males. About 2 kg/100 kg body weight of the pellet starter are required to achieve about 700 g ADG. A typical ration during this growing phase (average weight 125 kg) is composed of 2.5 kg of starter concentrate and 1 kg each of berseem hay and rice straw. Calculated intakes of nutrients in this ration are as follows: DM 4.1 kg, DM % of body weight 3.2, TDN 2.5 kg, DP 0.455 kg, ME/kg $W^{0.75}$ 229 kcal. Mudgal (1988) reported ME for maintaining buffalo heifers as 188 kcal/kg $W^{0.75}$, while Arora (1988) reported a value to 206 kcal/kg $W^{0.75}$ for maintenance and growth.

It has been shown that late-pregnant buffalo heifers need an extra 0.5 kg corn per day, in addition to the previously mentioned requirements. Aboul Ela (1988) concluded that resumption of cyclic activity post-partum in buffalo cows is influenced by feeding in late pregnancy and early lactation.

Feeding lactating buffalo cows

On a DM basis, buffalo's milk contains about 41% of total ingredients as fat and is thus characterized by a relatively high energy content, which should be carefully considered in the ration fed. A standard water buffalo cow weighing 500 kg, in her 3rd lactation, producing 7 kg/d milk for 300 days with average 7% fat requires 2 kg TDN and 400 g DP for maintenance plus 750 g TDN and 80 g DP per kg milk produced.

In Egypt, different concentrates and roughage ingredients are used to make up rations for lactating buffalo cows. Common concentrates which have been examined include cereal grains, cane molasses, cotton-seed cake, horse bean, soybean, linseed meal and sunflower seed. Common roughages include berseem hay; rice, wheat and barley straw; wheat and rice brans; rice hulls and water hyacinth hay or silage. The main green forage in winter is berseem (<u>Trifolium alexandrinum</u>) and its hay in summer.

In practice, a mixture of concentrates (60% TDN, 14% DP) is prepared in a cube form and comprizes yellow corn 23-25%, undecorticated cotton seed cakes 25-40%, wheat bran 10-15%, rice bran 10-15%, sugar cane molasses 3-6% and common salt plus lime stone 1.5%-2.5%. In winter, the feeding system of dairy buffaloes depends on green berseem and rice straw for dry, non-pregnant or early-pregnant buffalo cows. In summer, on the other hand, berseem hay replaces green berseem and green maize (darawa) is offered as a source of available vitamins (El-Ashry, 1988).

Several research trials were conducted to evaluate different roughages and concentrates for lactating buffalo cows. The ultimate goal of these trials was to introduce cheaper feed ingredients at maximum rate in the rations. The relatively cheaper roughages, mechanically-treated non-classical roughages like cotton stalks and corn cobs, were sprayed with sugar cane molasses to improve their utilization by lactating buffalo cows (El-Serafy, 1968). Although hazardous and expensive, NaOH treatment significantly improved the utilisation by lactating buffalo cows of poor quality roughages (rice and wheat straws and cotton stalks) (Abou Raya, 1967). The level of roughages in rations for lactating buffaloes has been generally accepted as 50% of total DM.

Research on the use of cheaper sources of nitrogen indicated that the urea can replace up to 50% of total nitrogen of rations for lactating buffaloes with no adverse effect on milk or fat yield (Khattab <u>et al</u>., 1981).

The level of concentrates in rations for milk production from buffaloes has a significant effect on milk and fat yields and the efficiency of dietary energy utilization (El-Ashry <u>et al</u>., 1975; the results are summarized in Table 2. Their data indicate that the efficiency of dietary utilization for milk production was significantly greater in winter than in summer. The level of 50% concentrates was more efficiently utilized for milk energy or protein but more than 60% concentrates in the ration reduced milk fat.

When intake is equal, the efficiency of utilization of the above feed ingredients for milk production is not different within a class of feedstuff. Mudgal (1988) discussed several factors affecting feed requirements and utilisation by buffalo cows and indicated that in dry subtropic regions temperature, shade, water requirement and disease are most important factors.

Table 2 Effect of level of concentrate and season on efficiency of dietary energy utilisation for milk production

Feeding season	Concentrate level %	Net efficiency of use of dietary energy %
Winter	0	65.2
	25	68.0
	50	67.6
Summer	0	60.2
	25	60.9
	50	62.0

Feeding bulls for production and draught

Feeding riverine buffalo bulls for production and draught is not common practice in Egypt. For doing some light work and for exercise, 5-6 year old breeding bulls are used to operate waterwheels ("sakia") for irrigation. The share of buffaloes in farm work was calculated as less than 8% of annual working hours regardless of farm size (Soliman, 1985). Requirements of breeding/draft bulls were calculated by Ranjhan & Pathak(1983).

OTHER MANAGEMENT ASPECTS IN REARING RIVERINE BUFFALOES

The following practices are used in improved management systems in Egypt:

- supplementary minerals and vitamins when animals are consuming dry feeds (summer feeding),
- spraying animals with water twice/d in summer (June-August),
- tethering fattened calves or lactating buffaloes and loose-housing for heifers,
- use of locally available materials for making sheds (the roof for sheds is made of rice straw sandwiched between two light-wood-framed bamboo mats),
- detecting oestrus with the bull twice daily,
- artificial insemination (using fresh semen) 10-12 hrs from the first natural mating,
- using mechanical milking machines, and
- regular (weekly) veterinary checks and assistance of the veterinarian in heifers delivering their first calf.

FUTURE OUTLOOK

Research should concentrate on biotechnology aspects such as super ovulation, embryo transfer, hormonal treatments to increase milk production and manipulation of rumen micro-organisms for better utilisation of ligno- cellulosic bonds in high-fibre-containing roughages. With increasing demand for milk and the noticeable decrease in the area of forage every year, there is an urgent need for cross breeding, possibly with a smaller but more productive strain. Otherwise buffaloes could loose ground to the efficient crossbreds from exotic cattle.

REFERENCES

Abou-Akkada, A.R. and El-Shazly, K., Effect of absence of ciliate
1966 protozoa from the rumen on microbial activity and growth of lambs. Applied Microbiology, 12: 384-389.

Aboul-Ela, M.B. Reproductive patterns and management in the buffalo.
1989 Proceedings of the International Symposium on the Constraints of Ruminant Production in the Dry Subtropics, Cairo, Egypt 5-7 Nov. 1988. pp. 174-185, EAAP Pub. No. 38.

Abou-Raya, A.K. Animal and poultry nutrition. 1st Edition.
1967 Dar-El-Maarif bookstore, Cairo, Egypt.

Afifi, Y.A. The effect of using different sources of roughages on
1977 carcass traits. Egypt. J. Anim. Prod. 17 (2): 165-171.

Arora, S.P. Management and feeding of buffalo calves from birth to
1988 early maturity. Proceedings of II World Buffalo Congress. New Delhi, India, 1988. Vol II, Part II, pp. 537-547.

Central Agency for Public Mobilization and Statistics. Statistical
1952-85 yearbook of ARE, June 1986.

Devendra, C. Forage supplements in feeding systems for buffaloes.
1987 Buffalo J. 2: 93-113.

El-Ashry, M.A. Impact of feeding and management on maturity in
1988 buffalo. Proceedings of II World Buffalo Congress. New Delhi, India, 1988. Vol II, Part II, pp. 548-555.

El-Ashry, M.A., El-Sherafy, A.M., Khattab, H.M., Aly Abbas, M. and
1981 Zaky, A.A. Effect of milk replacer diets on post-weaning performance and carcass measurements of buffalo calves. Indian J. Anim. Sci. 51 (4): 32-38.

El-Ashry, M.A., Shehata, O., Aboul-Selim, I.A. and Saleh Youssef, A.S.
1975 The effect of level of concentrates in the ration on the efficiency of feed utilization for milk production in buffaloes. Agric. Res. Rev. 53: 45-50.

El-Bassiony, A.Z. Productive performance of artificially reared
1983 buffalo calves. Ph.D. Thesis. Dept. Anim. Prod. Fac. Agric. Ain Shams Univ. Shoubra Al-Kheima, Cairo, Egypt. pp. 120-132.

El-Fouly, M.A. and Afifi, Y.A. The effect of age and weight at first
1977 calving on the productive and reproductive performance of primiparous buffaloes. Egypt. J. Anim. Prod. 17 (2): 117-124.

El-Koussy, H.A. Fattening buffalo calves fed on different sources and
1981 ratios of roughages and concentrates. Ph.D. Thesis. Dept. Anim. Prod. Fac. Agric. Ain Shams Univ. Shoubra Al-Kheima, Cairo, Egypt. pp. 60-73.

El-Serafy, A.M. and El-Ashry, M.A. The nutrition of Egyptian water
1989 buffaloes from birth to milk and meat production. Proceedings of the International Symposium on the Constraints of Ruminant Production in the Dry Subtropics, Cairo, Egypt 5-7 Nov. 1988. pp. 230-243, EAAP Pub. No. 38.

El-Serafy, A.M. Some nutritional studies on the suitable combinations
1968 of molasses with certain feeding-stuffs. M.Sc. Thesis. Dept. Anim. Prod. Fac. Agric. Cairo Univ. pp. 45-59.

El-Serafy, A.M., El-Ashry, M.A., Khattab, H.M., Soliman, H.S. and
1982 Zaky, A.A. Milk replacer diets for buffalo calves. 3. Economical evaluation of four formulas on the basis of nutrient availability and carcass yield. Indian J. Anim. Sci. 52 (10): 866-871.

Jasiorowski, H.A. Perspective and prospective of buffalo breeding for
1988 milk and meat production in the world. Proceedings of II World Buffalo Congress. New Delhi, India, 1988. Vol II, Part I, pp. 285-294.

Khattab, H.M., El-Ashry, M.A., El-Serafy, A.M. and El-Shoubokshy, A.S.
1981 Effect of feeding different levels of urea on the production performance of milking buffaloes. Egypt. J. Anim. Prod. 21 (2): 155-166.

Mudgal, V.O. Comparative efficiency for milk production of buffaloes
1988 and cattle in the tropics. Proceedings of II World Buffalo Congress. New Delhi, India, 1988. Vol II, Part II, pp. 454-462.

NRC. Nutrient requirements of dairy cattle. Fifth revised edition.
1978 Washington, DC.

Ragab, M.I., Darwish, M.Y. and Malek, A.G.A. Meat production from
1966 Egyptian buffalo. J. Anim. Prod. U.A.R. 6: 9-30, 31-50.

Ranjhan, S.K. Feeding strategies for the utilization of cellulosic
1988 wastes and other industrial by-products for meat and milk production in buffaloes. Proceedings of II World Buffalo Congress. New Delhi, India, 1988. Vol II, Part II, pp. 424-437.

Ranjhan, S.K. and Pathak, N.N. 1983. Management and feeding of
1983 buffaloes. Second edition. Vikas Pub. House. pp. 210-229.

Shehata, O., El-Ashry, M.A., Youssef, M.S. and El-Kholy, M.E.
1973 Fattening buffalo calves on rations containing different sources of roughages. Agric. Res. and Rev. Cairo. 51: 31-40.

Soliman, I. An analysis of the buffalo milk response under
1985 conventional Egyptian farming system. The 10th International Congress for Statistics, Computer Science, Social and Demographic Research. Cairo, Egypt, 1985. pp. 91-105.

Zaki El-Din, M., El-Shazly, K. and Hassouna, M. The relation between
1985 ruminal bacterial-N and diamino pimelic acid and bacterial activity on degrading this amino acid in farm animals. Egypt. J. Anim. Prod. 25: 297-309.

FEEDING SWAMP BUFFALO FOR MILK PRODUCTION

by

S. Khajarern and J.M. Khajarern

INTRODUCTION

In South-east Asia, the water buffaloes are predominantly of the swamp type. They are raised in small herds of 1-5 on small farms (2 to 5 hectares). Buffalo raising in this region cannot be considered as a distinct enterprise, rather it is an integral part of crop (mainly rice) production. Apart from serving as the main source of farm power, their integral roles in the small farm system include being the producers of farm manure, the removers of farm wastes, the living money savers (which act as the most reliable alternative source of farm income) and to exhibit the social status of the owners. Beef is no more than a by-product from buffaloes which are too old or not fit for work. Milk hardly plays any significant role in the small farm systems.

Traditionally, care of buffaloes is the responsibility of farm children. The animals are kept within the village, usually under the house, at night and spend the day-time hours grazing in the harvested paddy fields, along roadsides and on the edges of cultivated plots. In the rainy season, the animals are tethered on a small plot of land set aside for this purpose and supplemented with cut-and-carry grasses and/or rice straw. Minerals, except salt, are not generally offered. Water supply, for both wallowing and drinking, becomes a problem in many areas, particularly in Northeast Thailand during the dry season.

Breeding of village buffaloes is accomplished without human planning. It may occur at night or during the daytime when the animals are released and begin to mix with the village herd, as they do at the grazing or watering sites. This implies that mating does not generally occur during the rainy season when most of them are engaged in land preparation. This, along with other unfavourable factors such as the typical low reproductive rate of the species, high calf mortality and insufficient feed supplies, makes the herd productivity low and has led to a gradual decrease in size of the national herds in various countries (Bhannasiri, 1980; Toelihere, 1980). In order to cope with the increasing demands for draught power and beef, Frisch and Vercoe (1984) have outlined a strategy to increase both numbers and productivity of the herd in the region. Among these, ways of improving reproduction efficiency, draught output, growth rate and milk yield are discussed, taking into account the limitations set by the fact that the main feeds are made up of cereal straws and crop by products.

REGIONAL INTERESTS IN IMPROVING BUFFALO PRODUCTIVITY IN SMALL FARM SYSTEMS

The importance of buffalo as the indispensable support to crop production in small farm systems and the need to improve its productivity have been well realized by researchers and administrators of various institutions since the early seventies. A comprehensive overview, highlighting such interests, has been given by Soni (1985).

In Thailand, the Cooperative Buffalo Production Research Project, which was initiated in 1971 to join the inter-institutional efforts in solving two key problems of the decrease of buffalo number in the national herds and the reduction of mature body weight and size, has subsequently become the National Buffalo Research and Development Center, jointly undertaking research responsibility for several aspects for improving the buffalo productivity under small farm conditions.

The research thrust includes health care, reproductive physiology and artificial insemination, genetic improvement, nutrition and the integration of milking buffaloes into the farming system in some specific regions. The research results are regularly published in its annual reports and exchanged with its regional counterparts by means of regular publishing of "Buffalo Bulletin" under the support of the International Buffalo Information Center and the Regional Buffalo Development Network or through the "Buffalo Journal" of Chulalongkorn University. Recent technical information obtained, particularly on the digestion of fibrous residue feeds and on the feeding and management of buffaloes for milk production, will be highlighted in the following sections.

COMPARATIVE CHARACTERISTICS OF THE PHYSIOLOGY OF DIGESTION AND NUTRITION OF BUFFALO AND CATTLE

As with other ruminants, the buffalo has a remarkable ability to refine coarse roughages through rumen fermentation, leading to the formation and utilization of various essential metabolites for its nourishment. Anatomically, the rumen and reticulum of the buffalo are similar to those of cattle. However, the rumen of the buffalo, accounting for over 80 percent of the stomach capacity, is heavier than that of cattle and is 5-10 percent more capacious (Sengar and Singh, 1969). The buffalo omasum has lower tissue weight and capacity but the same number of laminae, having a narrower inter-laminar space than cattle. The abomasum of buffalo differs slightly in the distribution of cellular elements in the mucosa and its digestive ability is adversely affected by high air temperature than in the case of cattle (Chalmers, 1974). With respect to comparative physiology of digestion of buffalo and cattle, agreement has been reached in results from several studies in the particular aspects of:

1) the rumen of the buffalo calf becomes functional at an earlier age;
2) the microbial populations, bacteria and protozoa, are more numerous in the buffalo rumen;
3) the changes in the microbial populations, as affected by changes in season and thus proportions of food constituents, are more marked in the buffalo rumen;
4) the rate of passage of feed through the rumen of the buffalo is slower, allowing a longer retention time and exposure to more microbial action;
5) ammonia and soluble nitrogen disappear from the rumen fluid of buffalo more rapidly than from that of cattle, suggesting that the former utilizes protein more efficiently than the latter.

All of these characteristics tend to indicate that buffaloes have a higher efficiency of digestion than do the cattle. In his review of research results, derived from a large number of studies in the Indian subcontinent, Gupta (1988) concluded that buffaloes were more efficient converters of coarse roughages than cattle. His conclusion has been substantiated by the results of a number of trials which have shown higher concentrations of some metabolites in the rumen of the buffalo and higher digestibilities of dry matter, organic matter and crude fibre when compared to cattle. However, contradictory findings have also been regularly reported by a number of researchers such as Moran (1983). Chalmers (1974) expressed her doubts on such the claims due to the facts that :

1) a number of trials were involved with inadequate numbers of experimental animals;
2) studies should take into account differences in the kind of feeds, feeding levels, age and type of animals, rumen volume, rate of passage of feed, adequacy of water supply, deficiencies of vital feed constituents, management practices and so forth;
3) rumen content of various metabolites is always dynamic, with synthesis and breakdown occurring all the time. Variations in techniques used in sample preparation and/or chemical analysis would lead to different conclusions.

In her view, enough evidence indicates that there were only small differences in the efficiency of rumen digestion of fibre in cattle and buffaloes. However, she agrees that buffalo can utilize poor quality roughages more efficiently than cattle. A limited number of studies along these lines have been recently conducted with swamp buffaloes compared to Zebu cattle. Wanapat (1984) compared the dry matter degradability of 7 intact protein feeds by using the nylon bag technique on swamp buffalo and Brahman cross bulls being fed rice straw or urea-treated rice straw with 200 g/d fish meal. It was found that the DM degradability in buffalo was slightly higher than that in cattle, regardless of types of rice straw fed, at any 4-hour period ranging from 0 to 24 hours after suspension (Table 1).

Table 1. Dry matter degradability (%) of protein sources in the rumen of cattle and water buffalo.

Source	Duration (h.)									
	0		4		8		12		24	
	RS	UTS	RS	UTS	RS	UTS	RS	UTS	RS	UTS
Fish meal										
Buffalo	14.9	15.7	16.7	33.1	24.4	27.1	31.5	29.6	37.7	41.5
Cattle	15.5	16.3	26.3	29.0	25.3	27.1	29.4	27.9	36.0	39.6
Soybean meal										
Buffalo	23.5	33.9	29.4	44.5	34.2	31.3	33.8	38.5	66.7	66.1
Cattle	25.8	28.9	31.4	39.7	29.5	37.0	42.1	42.4	54.5	63.1
Leucaena leaf meal										
Buffalo	11.4	25.3	20.3	22.9	23.5	20.1	28.1	30.7	32.0	48.6
Cattle	9.0	23.3	26.5	26.1	23.5	21.9	26.0	35.2	32.5	44.9
Water hyacinth leaf meal										
Buffalo	24.9	27.6	21.4	27.5	27.5	23.3	28.7	32.4	29.9	41.7
Cattle	24.3	27.8	26.3	31.3	28.3	24.8	24.7	29.0	30.3	36.1
Cassava leaf meal										
Buffalo	11.4	21.8	16.9	20.5	19.2	19.2	24.6	27.5	34.0	43.1
Cattle	14.7	16.9	19.2	21.1	22.4	21.8	21.5	24.7	35.2	45.2
Sunnhemp leaf meal										
Buffalo	13.0	15.6	17.8	22.4	16.4	16.1	20.1	23.6	30.8	36.0
Cattle	10.2	21.9	13.1	20.1	20.1	18.4	21.0	22.1	24.5	27.7
Rice bran										
Buffalo	8.3	15.8	18.5	37.5	24.9	30.6	29.3	39.7	45.3	49.6
Cattle	8.7	10.8	18.2	28.5	26.4	28.4	32.4	33.7	40.3	47.5
Buffalo	15.3	22.2	19.5	29.8	24.3	24.1	28.0	31.7	39.5	46.7
	±6.4	±7.1	±5.0	±9.0	±5.8	±5.9	±4.5	±5.8	±13.9	±9.7
Cattle	15.5	20.8	23.0	28.0	25.1	25.6	27.9	30.7	36.2	43.5
	±7.1	±6.6	±6.3	±6.6	±3.3	±6.1	±7.3	±6.9	±9.4	±11.0

RS = rice straw, UTS = urea-treated rice straw

From Wanapat (1984)

The same group of workers (Chanthai et al., 1986) measured the metabolites in the rumen of a buffalo and a bullock being fed rice straw or urea-treated rice straw. It was shown that the rumen NH_3-N and rumen pH in the buffalo were higher than those of bullock (5.81 v. 4.49 mg % and 7.31 v. 7.01). The total VFA's were, however, significantly lower in buffalo than in cattle regardless of dietary treatments (Table 2).

Table 2. Comparison of NH_3-N, pH and total VFA between cattle and buffalo fed on rice straw (RS) and urea-treated rice straw (UTS).

Item	Feed		X±SEM
	RS	UTS	
NH_3-N, mg %			
Cattle	0.47	8.51	4.49±0.26^{a}
Buffalo	1.28	10.34	5.81±0.26^{b}
X±SEM	0.88±0.26^{a}	19.43±0.26^{b}	
pH			
Cattle	7.06	7.04	7.05±0.02^{a}
Buffalo	7.24	7.38	7.31±0.02^{b}
X±SEM	7.16±0.02^{a}	7.21±0.02^{a}	
Total VFA, mole/l			
Cattle	68.86	87.86	78.33±1.40^{a}
Buffalo	58.72	72.38	65.55±1.40^{b}
X±SEM	63.76±1.40^{a}	80.12±1.40^{b}	

a,b Value in the same row or column under appropriate headings with different superscripts differ ($P<0.05$).

From Chanthai et al. (1986)

With a higher quality roughage, Mahyuddin and Jalaludin (1986) compared the rumen metabolites and dry matter and nitrogen disappearances by nylon-bag techniques in 4 Kedah-Kelantan cattle and 4 swamp buffaloes being fed guinea grass (Panicum maximum) ad libitum with free access to mineral blocks. It was demonstrated that the rumen pH was lower in buffalo while the VFA's level was higher than those in cattle. The rumen NH_3-N levels were similar in both species and stayed at a low level after reaching a peak at 3 hours after feeding. The *in situ* degradability of DM and N of guinea grass showed an increasing trend throughout the first five 10-hour incubation periods and levelled off afterwards to 72 hours of incubation. The DM and N degradation rates of buffaloes (4.00±0.40, 3.90±0.55% h) were significantly higher than those of cattle (3.59±0.37, 2.83± 0.50% h).

On rice straw-based complete diets containing a 2% increment in crude protein ranging from 6 to 22 percent, Devendra (1985) compared the nitrogen utilization in 4 buffalo bulls and 5 Kedah Kelantan cattle by a balance trial. It was found that the buffaloes had higher N retention due to a significantly lower urinary nitrogen excretion than did the cattle. In both species, N intake was significantly correlated to apparently digestible nitrogen and N balance. The author stated that the DCP requirement for maintenance were 1.50 and 1.37 $g/W^{0.75}kg/day$ for the buffalo and cattle, respectively.

No firm conclusion can be drawn from the available data on whether or not buffalo can digest crude fibre more efficiently than cattle. The state of knowledge, however, strongly indicates that buffaloes can perform better on the poor quality roughages such as those that are available in small farms throughout Asia, namely the fibrous crop residues. In addition, the available data indicate that the buffalo utilizes protein more efficiently than cattle. This indication lends itself well to the possibility of improving buffalo productivity under the limited resources of small farms.

FEEDING AND MANAGEMENT OF BUFFALOES FOR MILK PRODUCTION

The swamp buffaloes are generally recognized as poor milkers when compared to the riverine breeds. The average milk yield ranges from 1.0 to 1.5 kilograms per head per day over a 270 to 305 days lactation (Castillo, 1978; Wejaratwimon *et al.*, 1979 and Thawinprawat *et al.*, 1985). Their potential for milk yield seems not to be greatly improved by improved feeding and management conditions (Frisch and Vercoe, 1984). However, they are of great value as the basic genetic stock from which animals with a greater potential for meat, milk and draught may be developed. Cross breeding of the swamp with the riverine buffaloes has been one of the ways to exploit the milking potential of the existing meat/draught animals in Thailand as well as in other Asian countries (Table 3). Tumwasorn (1981) conducted a comprehensive review on milk production and draught ability of the swamp-Murrah crossbred and reported that the average milk yield of the crossbred was 4 kg/d (2.0-6.3 kg/d) over the 256-day average lactation period. In his separate case study covering 6 herds of the Murrah crossbred cows in a village, he reported an average milk yield of 6.9 kg/d over a 242-day lactation. Working ability of the crossbred has been indirectly reported in terms of body weight and is not very meaningful.

Table 3. Milk yield of different buffalo breeds.

Breed	Lactation length (days)	Average daily milk yield (kg)	Note
Local (Swamp)[1]	236	1.94	Partly work
Murrah x local[1]	277	3.73	Partly work
(M x L) x Murrah[1]	292	5.20	
Murrah[1]	237	6.60	
Swamp x Murrah[2]	256	4.00	
Swamp x Murrah[2]	242	6.90	

[1]From Xiao (1988)
[2]From Tumwasorn (1981) - see previous page.

Konanta et al. (1984) compared the working ability of 16 Murrah crossbred with 16 swamp buffaloes being supplemented or not with 1.5 kg/d a 3:1 mixture of cassava chips and ipil-ipil leaf meal. It was demonstrated that the swamp buffaloes had a higher work ability in terms of being able to plough more land per unit of time but at a similar speed when compared to the crossbred. In addition, the groups that were fed the supplementary diet could plough at a significantly higher speed and gained more weight than the non-supplemented ones.

Not many studies on feeding and management of buffaloes for milk production have been conducted. Dairy farmers in the region generally take the conventional feeding system, good quality forages with concentrate feeds supplementation, for granted and the available resources may not always be used at their maximum economic efficiency. Taking into account the limited availability of good quality forages under small farm conditions, the low producing milking buffaloes of the multi-purpose crossbreds may have to produce on crop residue-based feeds. An alternative approach, enabling a more efficient utilization of such feeds, has to be formulated, tried out and implemented.

One of the alternative approaches to achieve a better utilization of fibrous crop residues as feeds for milking buffaloes is pre-treatment prior to feeding. A number of studies have been done, in South Asian countries, on evaluating the feeding value of urea-treated rice straw for dairy buffaloes.

Positive responses have been reported by Khan and Davis (1981) and Perdok *et al.* (1982 and 1984) (Table 4). Once such pre-treatment has proved socio-economically acceptable to small farmers, much benefit can be gained. This remains to be proven.

Table 4. The performance of Surti buffaloes given untreated or urea-treated straw-based diets, with concentrates and with or without Glyricidia leaves.

	UTS+C	UTS+G+C	TS+C	TS+G+C
Milk yield (kg/d)	2.17	2.56	2.97	3.35
Milk fat (g/d)	146	178	224	256
Milk fat (%)	6.71	6.94	7.54	7.62
Cows in milk after 10 wks	6	9	10	9
Wt gain of calves (g/d)	165	265	295	344
Wt change of cows (g/d)	-93	+59	+59	+126
DM intake (g/kg $W^{0.75}$)	119	123	163	178

UTS = untreated straw; TS = Treated straw; +C = + 1 kg concentrates/d
+G = + 6 kg Glyricidia leaves per day

From Perdok *et al*. (1982)

Supplementation of crop residue-based feeds with a more nutritious locally available by-products is another alternative for better utilization of the available feed resources in the small farm system. Preston (1986) and Preston and Sansoucy (1987) formulated a method of strategic supplementation in the feeding system by taking into account the ecological manipulation of rumen microbes so that the maximum digestion of dietary fibre can be achieved. In addition, supplementation of the by-pass nutrients required to correct the major constraints of milk production has been included in their recommendation. With respect to management, "restricted suckling" of the dual-purpose cow by her calf, enabling both calf growth and milk production to benefit, has been recommended. This appears to fit well with the small farm situation and remains to be tried out on South-east Asian small farms.

RECOMMENDATIONS FOR FUTURE RESEARCH AND DEVELOPMENT

In order to maximise the production potential of swamp buffaloes in South-east Asian small farms, work on crossbreeding with the dairy breeds is needed. Well-planned breeding programmes, aiming to produce multi-purpose crossbreds, should be done in order to make full use of the animals which are normally engaged in crop cultivation for only 130 days a year. Performance testing of the crossbreds needs to be undertaken on farms in order to simultaneously assess their suitability under the socio-economic setups of the small farmers.

A village survey which aims to assess basic data on the acceptance by small farmers of changing the pattern of raising the draught to multi-purpose buffaloes in their systems is greatly needed. Reorientation of their attitudes toward making use of their female buffaloes for both work and milk should be advocated.

On feeding and management, work on the assessment of crop residues and farm by-products available as feeds should be initially evaluated. The improvement of their utilization whether by means of pre-treatment and/or supplementation should be investigated and further tested on farms in order to formulate practicable and acceptable recommendations, enabling a sustainable milk production system based on the small farmers' available resources. A comparative study on the conventional rearing and restricted suckling of calves should be conducted in order to assess the economic gains, taking into account the improvement in both calf growth and milk yield.

REFERENCES

Bhannasiri, T. 1980 Buffalo production and development in Thailand. Buffalo Production for Small Farms. Book Series No. 15. Food and Fertilizer Technology Center, Taiwan. pp 21-29.

Castillo, C.S. 1975 Production, characteristics and processing of buffalo milk. The Asiatic Water Buffalo. Food and Fertilizer Technology Center, Taiwan. pp 36-58.

Chalmers, M.I. 1974 Nutrition. The Husbandry and Health of the Domestic Buffalo. Cockrill, W.R. (ed.). Food and Agriculture Organization of the United Nations, Rome, Italy. pp 167-194.

Chanthai, S., Wanapat, M., and Wachirapakorn, C. 1986 Rumen ammonia-N and volatile fatty acids concentrations in cattle and buffalo given rice straw based diets. 1986 Annual Report. The National Buffalo Research and Development Center. Kasetsart University, Bangkok, Thailand. pp 33-39.

Devendra, C. Comparative nitrogen utilization in Malaysian swamp
1985 buffaloes and Kedah-Kelantan cattle. Proceedings The 3rd AAAP Animal Science Congress. May 6- 10, 1985. Seoul, Korea. pp 873-875.

Frisch, J.E. and Vercoe, J.E. Improving the swamp buffalo of South-
1984 east Asia. Evaluation of Large Ruminants for the Tropics. Copland, J.W. (ed.). The Australian Centre for International Agricultural Research. Proceedings Series No. 5. Canberra, Australia. pp 37-43.

Gupta, B.N. Buffalo as an efficient converter of coarse roughages.
1988 Buffalo Production and Health. Indian Council of Agricultural Research, New Delhi, India. pp 120-129.

Khan, A.K.M.N. and Davis, C.H. Effect of treating paddy straw with
1981 ammonia on the performance of local and crossbred lactating cattle. Maximum Livestock Production from Minimum Land. Jackson, M.G., Dolberg, F., Davis, C.H., Haque, M. and Saadullah, M. (eds.). The 3rd Annual Seminar, Joydebpur, Bangladesh. February 2-5, 1981. pp 164-180.

Konanta, C., Limsakul, A., Treephan, V., Srisan, Y., Onwan, N. and
1984 Konanta, S. Working capability of Murrah x Swamp crossbred and Thai swamp buffaloes. 1984 Annual Report. The National Buffalo Research and Development Center. Kasetsart University, Bangkok, Thailand. pp 107-113.

Mahyuddin, N.A.M. and Jalaludin, S. Fermentation and degradation of
1986 guinea grass in cattle and buffaloes. Livestock Production and Diseases in the Tropics. Proceedings of 5th International Conference. University of Pertanian Malaysia, Selangor, Malaysia. August 18-22, 1986. pp 9-10.

Moran, J.B. Aspects of nitrogen utilization in Asiatic water buffalo
1983 and zebu cattle. J. Agric. Sci., Camb. 100: 13-23.

Perdok, H.B., Muttettuwegama, G.S., Kaesschieter, G.A., Boon, H.M.,
1984 van Wageningen, N.M., Arumugam, V., Linders, M.G.F. and Jayasuriya, M.C.N. Production responses of lactating or growing ruminants fed urea-ammonia treated paddy straw with or without supplements. Proceedings of 3rd Annual Workshop on The Utilization of Fibrous Agricultural Residues as Animal Feeds. School of Agriculture and Forestry. University of Melbourne, Victoria, Australia. pp 213-230.

Perdok, H.B., Thamotharam, M. Blom, J.J., vanden Born, H. and van
1982 Veluw, C. Practical experience with urea-ensiled straw in Sri Lanka. Maximum Livestock Production from Minimum Land. Preston, T.R., Davis, C.H. Dolberg, F., Haque, M., and Saadullah, M. (eds.). The 3rd Annual Seminar, Mymensingh, Bangladesh. February 15-18, 1982. pp 123.

Preston, T.R. Better Utilization of Crop Residues and By-products in
1986 Animal Feeding : Research Guidelines. 2. A practical manual for research workers. FAO Animal Production and Health Paper 50/2. FAO, Rome, Italy. 154 pp.

Preston, T.R. and Sansoucy, R. Matching animal production systems to
1987 available seed resources. Proceedings of the FAO Expert Consultation on the Substitution of Imported Concentrate Feeds in Animal Production Sustem in Developing Countries. FAO Animal Production and Health Paper 63. FAO, Rome, Italy. pp 32-49.

Sengar, O.P.S. and Singh, S.N. Studies on structure and development of
1969 compound stomach in buffalo (Bubalus bubalis). Proceedings on Ruminant Digestion and Metabolism. Indian Veterinary Research Institute. Izatnagar, India. December 15-23, 1969.

Soni, B. Buffalo research and development priorities for small farms
1985 in Asia. Proceedings of The Buffalo Seminar. Chantalakhana, C. (ed.). International Buffalo Information Center, Kasetsart University, Bangkok, Thailand. pp 35-44.

Thawinprawat, S., Na Chiangmai, A., Sirinantagate, K., Chatpongsri,
1985 C., Uriyapongsan, S. and Chaiyakarn, W. Preliminary study on yield and composition of swamp buffaloes' milk. 1985 Annual Report, The National Buffalo Research and Development Center, Kasetsart University, Bangkok, Thailand. pp 18-23.

Toelihere, M.R. Buffalo production and Development in Indonesia. Book
1980 Series No. 15. Food and Fertilizer Technology Center, Taiwan. pp 39-54.

Tumwasorn, S. Crossbreeding of swamp buffalo for milk and draft. Book
1981 Series No. 22. Food and Fertilizer Technology Center, Taiwan. pp 141-151.

Wanapat, M. Comparative dry matter degradation of intact leaf protein
1984 sources in fistulated rumen of cattle and buffalo fed untreated or urea-treated rice straw. 1984 Annual Report, The National Buffalo Research and Development Center, Kasetsart University, Bangkok, Thailand. pp 17-20.

Wejaratwimon, T., Leenanurak, D. and Mongkonpunya, K. Milk production
1979 and growth of swamp buffalo. Buffalo Bulletin Vol. 2. No. 2: 17-24.

FUTURE PROSPECTS FOR FODDER AND PASTURE PRODUCTION

By

A. Aminah and C.P. Chen

INTRODUCTION

In the wet tropical environment, there is no distinct seasonal moisture deficiency and the foliage is green throughout the year. The potential production of tropical forage both native and improved species, in terms of protein, metabolisable energy and milk production, has been favourably assessed (Lane and Mustapha, 1983; Luxton, 1983).

The tropical dairy breed, which is basically a <u>Bos indicus</u> animal, has a low potential for milk production which is between 500-950 kg/lactation (Samuel, 1974; Sivarajasingam, 1974), whereas the milk production potential of the crossbred (<u>Bos taurus</u> x <u>Bos indicus</u>) is improved, giving 1,200-1,900 kg/lactation (Sivasupramaniam and Nik Mahmood 1981). Due to the overall feed constraint and environmental stress, purebred dairy cows such as Friesian and Jersey are merely able to produce half of their milking potential (1,150-2,200 kg/lactation) (Wan Hassan <u>et al</u>., 1981; Sivasupramaniam and Nik Mahmood, 1981), compared to those of similar breeds in a drier environment. The expected milk yield is estimated to be 2,700-4,000 kg/lactation for the Friesian and Jersey (Cowan <u>et al</u>., 1975). Hence, the genetic expression of milk production of the dairy cow is confounded by tropical environments.

This paper attempts to present the status, constraints, potential production and management of tropical forages in relation to the feeding of dairy cattle in the tropics.

NUTRITIONAL VALUE OF TROPICAL PASTURES

It is commonly believed that the contribution in milk production due to genetic factors of animal is about 25%. A greater sustainability of production is produced by better feeding. Pasture can be a major source of feed for dairy cows but there are some limitations to its use. Energy and protein supplies are the most essential components in animal nutrition and, in many tropical countries, these components are often the critical limiting factors to animal production.

Most of the tropical grasses (either native or improved pastures) have metabolisable energy values ranging from 7.0 to 11.0 MJ/kg DM when cut between 2-8 weeks (Table 1), and energy concentrations for natural forages were found to be similar (7.1 to 10.1 MJ ME/kg DM). Lane and Mustapha (1983) also reported that broadleaved species and ferns appeared to have higher metabolisable energy values and their crude protein and crude fibre was superior to natural grasses. Based solely on the metabolisable energy available from the existing forage on offer, Lane and Mustapha (1983) estimated the potential milk production of Friesian-Sahiwal cows in mid-lactation to be 12-16 kg/cow/day depending upon the types of pastures.

Table 1 . Mean chemical composition and estimated ME value of some improved tropical pastures (Wan Hassan, 1987)

Grass/Cutting Interval (week)	*in vitro* DMD %	CP	CF	Ash	Estimated ME (MJ/kg DM)*
		g/kg DM			
Setaria					
2	66.00	189.86	256.69	99.61	9.18
4	60.29	143.17	286.17	98.58	8.42
6	56.09	121.95	314.42	86.38	7.96
8	53.09	104.71	329.27	77.67	7.62
Digitaria					
2	64.24	158.37	273.97	95.77	8.98
4	59.06	139.64	301.79	81.83	8.41
6	55.32	100.89	319.74	79.29	7.92
8	51.90	88.14	334.47	67.80	7.54
Napier					
2	65.60	178.08	256.54	115.55	8.97
4	59.35	124.31	301.65	95.55	8.32
6	54.60	100.00	330.58	91.00	7.71
8	50.92	70.15	343.20	73.33	7.36
Guinea					
2	64.04	172.87	279.34	111.23	8.80
4	57.69	127.27	324.10	88.17	8.16
6	53.88	91.05	350.90	80.63	7.71
8	49.89	62.57	367.36	69.83	7.24
Signal					
2	63.52	154.92	269.66	97.76	8.96
4	57.66	119.29	300.10	81.00	8.22
6	55.03	92.58	325.61	77.86	7.89
8	53.56	76.00	345.73	68.73	7.76

*ME (MJ/kg DM) = 0.15 (DMD%+2% units)$\left(\frac{100-\text{Ash}\%}{100}\right)$

Protein content varies with age, part of the plant and species. Most of the tropical pastures have crude protein contents ranging from 7 to 12% for grasses and more for legumes like Leucaena, which has 25% protein content. Protein content of tropical pastures decreases rapidly as growth progresses.

As a general guide, 10% crude protein on a dry matter basis is adequate for fattening cattle but about 15% crude protein or more is required for high producing milking cows. The critical level of crude protein required in the pasture before intake is reduced by a protein deficiency has been estimated at between 6.0 and 8.5% (Milford and Minson, 1966; Minson, 1967). The deficiency of crude protein in pasture can be corrected by the use of tropical legumes or nitrogen fertiliser on pure grass pastures.

The digestibility of cultivated tropical grasses lies between 50 and 65%, whereas temperate grasses range from 65 to 80% (De Gues, 1977). Thus it is necessary to utilise immature herbage in order to obtain a high metabolisable energy intake. Milford and Minson (1966) showed that the decline in digestibility with age was more rapid in tropical grasses compared with tropical legumes, which retained relatively high digestibilities at maturity. Values recorded for a number of different tropical grasses indicate that there is a decrease of 0.1 to 0.2 digestibility units/day with increasing maturity (Minson, 1971). This explains why tropical legumes are particularly valuable for animals in the dry season.

PASTURE SPECIES SELECTION AND PRODUCTION

Species Performance and Adaptation

Selection of pasture species should take into account the nutritional requirements of different classes of livestock, as well as the suitability of the plant for different animal production systems such as large scale or smallholder production.

Lately, improved species which had gone through the screening process in the wet tropics are the following genera:

Grass: *Brachiaria*, *Digitaria*, *Panicum*, *Setaria* and *Pennisetum*

Legume: *Centrosema pubescens*, *Desmodium ovalifolium*, *Pueraria phaseoloides* and *Leucaena leucocephala* (Wong *et al*., 1982).

The performance of the above mentioned species evaluated under different soil types are listed in Table 2. In general, on sedentary soil and peat soil, dry matter production ranged from 15.0 to 30.0 ton/ha/yr, whereas the same species grown on coastal marine sand dropped to almost a third or a half of their normal yields.

Table 2. Dry matter yield (ton/ha) of grasses and legumes under cutting in three regions in Malaysia

Species	Sedentary soil	Peat soil	Sandy soil
Grasses:			
Brachiaria decumbens	24.7	26.3	16.5
Brachiaria brizantha	19.4	24.5	11.8
Digitaria setivalva	20.5	25.4	3.8
Digitaria pentzii (Slenderstem)	18.2	23.1	11.8
Panicum maximum (Coloniao)	17.0	20.2	3.1
Panicum maximum (Typica)	26.1	-	-
Setaria sphacelata (Kazungula)	20.6	15.8	6.7
Setaria sphacelata (Splendida)	18.6	16.6	-
Pennisetum purpuruem	30.0	16.3	3.4
Legumes:			
Centrosema pubescens	6.0-10.0	5.0- 8.0	3.0
Desmodium ovalifolium	5.0- 7.0	7.0- 9.0	3.0
Stylosanthes guianensis(Schofield)	7.0-15.0	5.0- 7.0	7.0-10.0
Leucaena leucocephala	8.0-20.0	10.0-15.0	5.0- 6.0

Adapted from Wong et al., (1982).

Fertiliser applied at 300-400 kg N/ha/yr; cutting interval of 4-6 weekly.

The grass species Napier, and to certain extent Guinea, are mostly used for cutting while the rest are meant for grazing. In view of the grass-legume combining ability, it is proposed that the grasses with erect and clumpy characteristics such as Panicum maximum, Setaria sphacelata can combine well with all the promising tropical legumes, especially Desmodium ovalifolium and Centrosema pubescens, which need a shady canopy. Sometimes, legume species which are not that palatable, such as Calopogonium caerulum, C. mucunoides and S. scabra, may have to be included in the pasture system so as to serve specific objectives such as preservation of feed for drought and nitrogen fixation for soil improvement. Care is required with aggressive grasses with prostrate rhizomes which form a thick mat on the ground surface which may impede the legume. Selection of shrubby legumes for this system could be probably the best way out, for instance, the L. leucocephala and B. decumbens pastures.

Cutting and grazing

It is important to note that for most grasses and legumes, forage yield increases as cutting frequency decreases while forage quality declines. The digestibility of both grasses and legumes decreases with maturity, implying that forage should be fed at a younger stage for maximum energy digestibility. A wide range of digestibility occurs both between and within pasture species. One has to compromise between maximising forage yield and quality and try to improve the latter by using better species for milk production.

In general, defoliation affects the both above ground growth and the underground rooting system. In the case of legumes, it affects also nodulation and nitrogenase activity. When sufficient fertilizer and moisture are available, a 6 to 10 weeks regrowth interval should be the practice to obtain optimal yield and quality of forage, except in the dry season, when the cutting interval may inevitably be prolonged.

In the case of grazed pastures with defoliation by dairy cows, either by continuous or rotational grazing, the optimal leaf to stem ratio should be maintained at close to 1, giving forage availability of about 2500 kg/ha of dry matter at any time (Cowan et al., 1977)

Response to Fertiliser

In tropical regions, where light and moisture are non-limiting, soil nutrients are the major factors affecting the production of forage. Due to highly weathered soil conditions, deficiencies of macro- and micro-nutrients in ultisol, oxisol, peat and marine sand were reported (Coulter 1972; Chew et al., 1981; Tham and Kerridge 1979, 1982). Nutrient deficiencies may lead to the non-persistence of the species, especially with legumes which are sensitive to molybdenum, copper, magnesium, boron and calcium. They may eventually affect animal production, e.g. cobalt deficiency both in soil and pasture (Mannnetje et al., 1976b). There is a response of pasture growth and animal performance to the application of phosphorus fertilizer (Eng et al., 1978).

A high rate of nitrogen fertilizer is necessary to maintain high productivity of fodder grasses. The dry matter yields of some of the improved and native species in response to nitrogen fertilizer are documented (Ng, 1972, Vincente-Chandler et al., 1959. Dry matter yield responses have been recorded up to as high as 1,600 kg N/ha/yr (Tham, 1980). Even though high rates of nitrogen increase dry matter yield, the efficiency of use of nitrogen was found to decrease with increase rates of nitrogen applied. The nitrogen efficiency drops from 23.0 to 20.1 to 17.6 to 16.8 kg DM/kg N as nitrogen rate increases from 200 to 400 to 600 to 800 kg N/ha/yr respectively. Whereas at the same rate of application, the nitrogen recoveries were 30.3, 38.4, 41.9 and 42.7% (Chadhokar, 1978).

Similar results on napier and signal grass were recorded at rates of 42.0, 34.2 and 25.2% nitrogen recovery as the nitrogen application rate increases from 200-400, 400-800 and 800-1600 kg N/ha/yr (Tham,

1980; Aminah et al., 1989). It implies that the most efficient nitrogen fertilizer rate should be around the level of 200-400 kg N/ha/yr. This further confirmed an earlier finding that nitrogen concentration in the forage had a limited effect on increasing the nutritional value (Minson 1973) and that nitrogen fertilizer at 250 kg N/ha/yr was sufficient for the attainment of crude protein for optimum digestion by the animal in the wet tropics (Mustapha et al., 1987). Furthermore, excessive crude protein in tropical grasses following heavy application of fertilizer nitrogen may also adversely affect intake. Milford (1960) recorded a depression of 33% in the intake of *Chloris gayana* when crude protein content increased from 8 to 13.5% due to high nitrogen fertilizer application. A similar case was reported that the intake of young, heavily fertilised pasture with 20% crude protein was 28% less than the intake for the same pasture after growing for a further 28 days when the crude protein had fallen to 11%.

There are interaction of N, P and K fertilizers on forage production. With application of N, P and K fertilizer at the rates of 421, 196 and 1004 kg/ha/yr respectively, the yield of fresh napier grass increased by 74.5% over the yield of unfertilised grass. It is recommended for a broad range of soils in the humid tropics that fertilizer rates of 300-600 kg N, 100 kg P and 50 kg K/ha/yr should be sufficient for forage production under cut and carry system (Robbins, 1986). Based on research work and experience on grazed pastures, Gilbert (1984) has attempted to recommend fertilizer for different soils derived from granitic, metamorphic, basaltic and marine sand (Table 3).

Table 3. The fertilizer rates for grazed pastures on different soil types (Gilbert, 1984).

Fertilizer	Granitic soil	Metamorphic soil	Basaltic salts	Marine sand
Superphosphate	300 kg/2yrs	300 kg/2yrs	300 kg/yr	150 kg/yr
Muriate of Potash	100 kg/4-5yr	100 kg/4-5yr	-	50 kg/yr
Copper sulphate	8 kg/4yr	-	-	8 kg/4yr
Zinc sulphate	8 kg/4yr	-	-	8 kg/4yr
Sodium molybdate	0.5 kg/4yr	0.5 kg/4yr	0.5 kg/2yr	0.5 kg/4yr
Lime	-	500 kg/ha	-	-

On a pure grass sward grazed by dairy cattle, it is advisable to split the amount of 300 kg N/ha/yr of nitrogen fertilizer into five equal applications. The economic aspects of fertilizer use has to be assessed in relation to the increased dry matter yield and its subsequent effect on animal carrying capacity.

Legume in the Pasture

To maintain high productivity and forage quality for dairy cows, it may be better to include leguminous species in the pasture production system rather than rely on nitrogen fertilizer. The advantages of legumes in the system are:

(i) improvement of soil conditions due to nitrogen built up in the soil from accumulation of organic matter,

(ii) fixation of nitrogen by the legume through *Rhizobium* symbiosis, and

(iii) increased animal production due to the higher nutritive value of legumes and shorter digestive passage time in the gut that enhance voluntary intake.

Usually, the crude protein content of legumes (at about 25%) is higher than that of grasses at similar ages and stages of growth and shows little fluctuation during the growing process. Apart from higher nitrogen content, tropical legumes generally maintain higher sulphur (0.07-0.21%) and calcium (1.13-1.93%) in the plant tops (Andrew and Robbins, 1969) as compared to that of grasses (0.09-0.15% and 0.17-0.41%, respectively). Similarly, the values of phosphorus in legumes are expected to be higher than grasses despite great variability between species and plant age. The additional role of legumes in increasing the mineral content of pastures has an additive effect on animal nutrition and production.

Legume viability and persistence on acidic soil are always problematic. The inclusion of local legumes or well-adapted species as one of the leguminous components in the pasture is highly recommended. Because of specific requirements for certain elements for nitrogen fixation, the legume may be more sensitive to some minerals such as phosphorus, sulphur, calcium, magnesium, molybdenum and cobalt.

There are many factors affecting legume component in the tropical pastures. The most important but controllable factors are the defoliation, grazing management and fertilizer. Very little is known about the optimum level of legume content in the tropical pasture. Results from grazing pangola-legume pasture have shown that the live weight gain of beef cattle was linearly related to legume content of the pasture (Evans, 1970).

PASTURE MANAGEMENT AND MILK PRODUCTION

Average production per cow from tropical pastures is in the range of 10 to 12 kg/day for Friesian cows, 7 to 9 kg/day for Jersey and 6 to 10 kg/day for crossbred cattle. The potential of these pastures for milk production was suggested to be 4,000 kg/lactation for Friesian and 2,700 kg for Jersey (Cowan _et al_., 1974). Production per hectare varies from 2,600 to 8,300 kg/ha/yr. Grass and legume systems have produced up to 8,000 kg/ha/yr, but these stocking rates caused degradation of the pasture due to loss of legume (Cowan _et al_., 1975). A production level of 5,000 kg/ha/yr is the potential of stable grass and legume mixed pastures. (c)

Continuous Grazing

Cows grazing tropical pastures require about 10 to 12 hours a day of grazing to satisfy their nutritional needs (Cowan, 1975). Often they are reluctant to do a lot of grazing during the day as the temperature are high. In Northern Queensland during summer, about 80% of grazing are done during the night. Obviously, night grazing or feeding in the tropical environment, must be encouraged.

There is linear increase in milk output per hectare with increase in stocking rates but milk production per cow also decreases linearly (Cowan, 1984). There are a wide range of stocking rates tested for milking cows on different types of pastures (Table 4). The optimal stocking rate for Friesian milking cows on Guinea-Glycine mixed pasture was 1.6 cows/ha to produce 5,351 kg/ha/yr (or 3,345 kg/cow/yr) of milk, whereas on nitrogen fertilised Guinea was 3.5 cows/ha to produce 8,880 kg/ha/yr milk yield. Both stocking rates were able to maintain stable pasture of about 2,500 kg DM/ha on offer, the amount of forage considered to be minimum for dairy production.

For practical purposes, when advising a farmer on the basis of experimental results, it is best to be conservative at 20-30% lower stocking rates than those used in the research.

Rotational Grazing

On reviewing over 16 grazing experiments, Mannetje _et al_., (1976a) concluded that there is no definite advantage of rotational grazing over continuous grazing system. However, in the hot humid tropics where even rainfall is available, the rotational grazing system has a 25% higher in beef production than that of the continuous (Chen and Othman, 1986). This is attributed to a higher amount of forage on offer rather than forage quality in the rotational grazing system. The practice of rotational grazing (or strip grazing) may help to ease the forage problem, particularly during a prolonged dry spell.

Table 5. Milk production from N-fertilised and legume-based tropical pastures without supplementary feed

Pasture	Stocking Rate	Breed	Milk yield		References
			kg/cow /day	kg/ha/yr	
<u>Unfertilised pastures</u>					
<u>P. maximum/ M. minutiflora</u>	1.1	Jersey	6.8	2667	Toledo 1968
<u>D. decumbens</u>	1.5	Friesian/Zebu	6.9	3760*	Serpa *et al*., 1973
<u>Nitrogen fertilized pastures</u>					
<u>D. decumbens</u>	2.5	Jersey	6.8	6014	Toledo 1973
<u>D. decumbens</u> (irrigated)	8.9	Jersey	6.5	22466	Thurbon *et al*., 1973
<u>P. maximum</u>	2.5	Holstein	11.3	8488	Vincente-Chandler *et al*., 1974
<u>Grass-legume pastures</u>					
<u>B. decumbens/ Leucaena</u>	5.0	Sahiwal-Friesian	5.7	8580	Wong *et al*., 1987
<u>P. maximum/ Glycine</u>	2.5	Friesian	12.7	8221	Cowan *et al*., 1975

*Calculated yield

Cut and Carry System

The "cut and carry" (or zero grazing) system means that the fodder is cut and removed for stall-feeding to animals. This system has been widely adopted by smallholders in dairy farming. The reasons for this practice are the shortage of land, small scale of farm size (0.3 -2.0 ha), abundance of cheap labour, limited forage resources and strict control of animals.

Usually, when cut forage is given, the nutritive value of forage is inferior to that received by grazing animals. The protein content of napier grass by cattle was 17.1%, while that of the same forage being fed to stall kept animal was 7.4% (Vincente-Chandler et al., 1964). Grazing animals are able to choose their own forages. Grazing cows tend to produce more milk and obtain better reproductive performance than stall-fed cows. Milk production in the stall-feeding system was 8,577 kg/ha/lactation while rotational grazing without supplements yielded 9,180 kg/ha (Wong et al., 1987). A similar finding was obtained with 20% higher milk production for grazed cows (10,203 kg/ha/yr) than that of stall-fed cows (8,134 kg/kg/yr) (Soetrisno et al., 1985). This may explain the low milk yield of the smallholders.

Like grazing animals, stall-fed milking cows need night feeding. In order to have a near-balanced diet, some broadleaved weeds such as (*Asystasia intrusa* and others) or leguminous shrubs (such as *Leucaena* and *Glyricidia* species) in place of a high protein supplement, should be included. Protein is an expensive supplement at smallholder level.

Frequent defoliation of herbage is a threat to the persistence of sward and it may result in the need to replant the pasture. If it is a machine-cut, the mortality of forage plants will be higher than hand-cut materials. Experience shows that Napier grass cut by forage harvester has to be replanted every three years. The severity of forage die-back may be reduced if a reciprocal cutting machine is used.

Replenishment of soil fertility is essential with the cut and carry system. It was estimated that to be able to produce 150 t/ha/yr of fresh Napier/Guinea fodder, a fertilizer programme of 880 kg N, 252 kg P and 756 kg K/ha/yr (or 6.3 ton of 14:14:12 compound fertilizer) is needed. On the highly-leached acidic soils of the tropics, "soil exhaustion" may be experienced, despite the frequent application of N, P, K fertilizers. The consequence is the rapid die-back or retarded growth of forage sward. It may be due to the following factors:

(i) high hydrogen ion concentration (urea source),
(ii) toxic level of aluminium and manganese, or
(iii) induced deficiency of molybdenum.

Correction of such a problem probably requires the incorporation of organic matter or animal waste into the soil, as well as the application of recommended fertilizers. Organic matter which is the source of much nitrogen, and to a certain extent phosphorus and sulphur, improves soil condition. Organic matter improves also the inorganic particles (the clay colloids) which are the main reservoir of cationic nutrients such as K, Ca, mg, Fe, Zn, Mn and Co.

Some micro-elements which are not essential to plant growth such as sodium and cobalt are critical to animal performance. Even the major nutrients such as phosphorus, potassium, calcium and nitrogen if they are low in fodder plants, will affect milk production. Supplementation of minerals to animal to correct the immediate need must be given.

Forage-based Supplementary Feeding

In areas where dairy cattle are kept near a pineapple factory, palm oil mill, sugar-cane plantation or in any major agriculture operation from which crop by-products are plentiful, a complete year round feeding system involving these by-products could be established. By-products do not provide a balanced feed on their own, being either excessive or deficient in certain minerals, but they are good roughage with high metabolisable energy. They may be available or harvested only in a certain period of the year, mostly approaching the dry season when shortage of green forage is experienced.

Besides the agriculture by-products, the conservation of feed in the form of silage or hay may be another alternative to ease the situation of feed shortage on the farm. Due to low leaf:stem ratio or high fibre content, tropical pasture may not be that good for silage and hay making, but with some additional mixing with leguminous shrubs, good quality feed can still be maintained. Hence, using forage as a basic diet, supplementary feeding of formulated by-products in the ration, may possibly be able to maintain full milk production throughout the year.

REFERENCES

Aminah, A., Wan Mohamad, W.O. and Zulkifli, S. 1989. Seasonal Dry Matter Production and Nitrogen Fixation of Leucaena (*Leucaena leucocephala*) and Stylo (*Stylosanthes guianensis*) in Pure Swards and in Association with Signal Grass (*Brachiaria decumbens*). M. Agric. Sc. Thesis, UPM, Serdang Selangor.

Andrew, C.S. and Robbins, M.F. 1969. The effect of phosphorus on the growth and chemical composition of some tropical legumes. I. Growth and critical percentages of phosphorus. *Aust. J. Agric. Res.* 20:665-674.

Chadhokar, P.A. 1978. Effect of rate and frequency of nitrogen application on dry matter yield and nitrogen concentration of para grass (*Brachiaria mutica*). *Trop. Grassl.* 12:127-132.

Chen, C.P. and Othman, O. 1986. Effects of stocking rate and grazing management on cattle production from tropical grass, *Digitaria setivalva* Stent. *MARDI Res. Bull.* 14(1):69-80.

Chew, W.Y., Chin, Y.K., Ismail, L. and Ramli, K. Relative
1981 availability to plants of phosphorus from rock phosphate and triple superphosphate in relation to liming on Malaysian peat. MARDI Res. Bull. 9:49-57.

Coulter, J.K. Soil of Malaysia - A review of investigation on their
1972 fertility and management. Soils and Fertilizers, 35:475-498.

Cowan, R.T. Grazing time and pattern of grazing of Friesian cows on a
1975 tropical grass-legume pasture. Aust. J. Expt. Agric. Anim. Husb. 15:32-37.

Cowan, R.T. Responses to supplementation under tropical pasture
1984 feeding system. Fifth regional pastures training course. Pasture development and production in the high rainfall tropics of South-East Asia, 26 Nov-14 Dec, 1984 at UPM:FAO. 236-239.

Cowan, R.T., Byford, I.J.R. and Stobbs, T.H. Effect of stocking rate
1975 and energy supplementation on milk production from tropical grass-legume pastures. Aust. J. Exp. Agric. Anim. Husb. 15:740-746.

Cowan, R.T., Davison, T.M. and O'Grady, P. Influence of level of
1977 concentrate feeding on milk production and pasture utilisation by Friesian cows grazing tropical grass-legume pasture. Aust. J. Exp. Agric. Anim. Husb. 17:373.

Cowan, R.T., O'Grady, P., Moss R.J. and Byford, I.J.R. Milk and fat
1974 yields of Jersey and Friesian cows grazing tropical grass-legume pastures. Trop. Grassl. 8:117-120

De Gues, J.G. Production Potentialities of Pastures in the Tropics
1977 and Subtropics. Zurich: Centre d'Etude de l'Azote, pp 53.

Eng, P.K, Mannetje, L't. and Chen, C.P. Effects of phosphorus and
1978 stocking rate on pasture and animal production from a guinea grass-legume pasture in Johore, Malaysia. II. Animal liveweight changes. Trop. Grassl. 12:198-207.

Evans, T.R. Some factors affecting beef production from subtropical
1970 pastures in the coastal lowlands of South East Queensland. Proc. 11th Int. Grassl. Congr. Surfers Paradise. 1970.p 803-807.

Gilbert, M.A. Fertilizers and fertilizer application methods. Fifth
1984 regional pastures training course. Pasture development and production in the high rainfall tropics of South East Asia, 26 Nov-14 Dec, 1984 at UPM:FAO. p. 146-149

Lane, I.R. and Mustapha, M. Utilisation of natural pastures for dairy
1983 cattle in humid tropics. Paper presented to Vth World Conference on Animal Production, 14-19th august, 1983, Tokyo, Japan pp 2.

Luxton, M.J.F. Large scale dairy farming. Proc. 7th Ann. Conf.
1983 Malaysian Society of Animal Production 206-213.

Mannetje, L.'t., Jones, R.T. and Stubbs, J.H. Pasture evaluation by
1976a grazing experiments. In Tropical Pasture Research (ed. Shaw, N.H. and Bryan, W.W.) pub. by CAAAB Bull No. 51 Chap. 9 pp 202-210.

Mannetje, L.'t., Ajit, S.S. and Murugaiyah, M. Cobalt deficiency in
1976b cattle in Johore. Liveweight changes and response to treatments. MARDI Res. Bull. 4: 90-98.

Milford, R. Nutritional value of subtropical pasture species under
1960 Australian conditions. Proc. 8th Int. Grassl. Congr. Reading: p. 474.

Milford, R. and Minson, D.J. Intake of tropical pasture species.
1966 Proc. 9th Int. Grassld. Congr. Sao Paulo, pp 815- 822.

Minson, D.J. The voluntary intake and digestibility in sheep, of
1967 chopped and pelletad Digitaria decumbens (Pangola grass) following a late application of fertilizer nitrogen. Brit. J. Nut. 21: 587-597.

Minson, D.J. The nutritive value of tropical pastures. J. Aust. Inst.
1971 Agric. Sci. 22:255-263.

Minson, D.J. Effect of fertilizer nitrogen on digestibility and
1973 voluntary intake of Chloris gayana, Digitaria decumbens and Pennisetum clandestinum. Aust. J. Expt. Agric. Anim. Husb. 13:153-157.

Mustapha, M., Samuel, C. and Vijchulata, P. Effects of varying
1987 nitrogen levels in Setaria sphacelata var. Splendida on the efficiency of nitrogen utilisation in dairy bulls. Proc. 10th Ann. Conf. Malaysian Society of Animal Production p 152-156.

Ng, T.T. Comparative responses of some tropical grasses to fertilizer
1972 nitrogen in Sarawak, E. Malaysia. Trop. Grassl. 6:229-236.

Robbins, G.B. Fertilizer requirements of grasses in cut and carry
1986 system. Six Regional Training Course on Tropical Pasture Production for the High Rainfall Tropics of S.E. Asia. pp. 247-249.

Samuel, C. Milk yield potential of Local Indian Dairy cattle and
1974 their crossbreds at University Pertanian Malaysia. Mal. Vet. J. 5:79-88.

Sivarajasingam, S. The Red sindhi cattle of Malaysia 19th Annual
1974 Malaysian Veterinary association Conference, 26-27th October, 1974, Penang.

Sivasupramaniam, G. and Nik Mahmood, M. Comparison of performance of
1981 Friesian and Friesian cross LID cattle on Majuternak farms. Proc. 11th Ann. Conf. Malaysian Society of Animal Production. pp. 202-213.

Soetrisno, D. Mahyuddin, M., Vijchulata, P. Osman, A. and Tan, H.S.
1985 Performance of sahiwal-Friesian cows under grazing and stallfeeding management. Proc. 9th Ann. Conf. Malaysian Society of Animal Production. 65-71.

Tham, K.C. Nitrogen response of Napier grass (Pennisetum purpureum)
1980 on an ultisol formed from sandstone in Peninsular Malaysia. MARDI Res. Bull. 8:49-60.

Tham, K.C. and Kerridge, P.C. Nutrient responded by Stylosanthes
1979 guianensis on coastal marine sands (Bris) of Peninsular Malaysia. MARDI Res. Bull. 7: 43-54.

Tham, K.C. and Kerridge, P.C. Response to lime, K, Mo and Cu by
1982 grass-legume pasture on some ultisols and oxisols of Peninsular Malaysia. MARDI Res. Bull. 10:350-369

Vincente-Chandler, J., Silva, S. and Figarella, J. The effects of
1959 nitrogen fertilisation and frequency of cutting on the yield of I. Napier, II. Guinea and III. Para grass. J. Agric. Univ. Puerto Rico, 43:215-248.

Vincente-Chandler, J., Caro-Costas, R., Pearson, R.W. Abruna, F.
1964 Figarella, J. and Silva, S. Intensive management of tropical forage in Puerto Rico. Univ. Puerto Rico Agric. Expt. Sta. Bull. 187.

Wan Hassan, W.E., Abu Bakar, C. and Tan, K.L. The performance of
1981 imported Jersey cattle in Malaysia I: Milk production. Proc. 5th Ann. Conf. Malaysian Society of Animal Production, 192-201.

Wong, C.C., Chen, C.P. and Ajit, S.S. A report on pasture and fodder
1982 introductions in MARDI. MARDI Report. No. 76, 35 pp.

Wong, C.C., Wan Hassan, W.E. and Tan, K.L. Milk production of Sahiwal
1987 x Friesian cow grazing on leucaena-B. decumbens pasture with concentrate supplementation. Proc. 10th Ann. Conf. Malaysian Society of Animal Production. 280-286

FORAGE AND LEGUMES AS PROTEIN SUPPLEMENTS FOR PASTURE BASED SYSTEMS

by

F.A. Moog

INTRODUCTION

Major emphasis for pasture improvement in the tropics has been placed on grass-legume pastures. The grass-legume approach in pasture development is based on the knowledge that tropical soils often lack nitrogen and on the general philosophy that the legume-*Rhizobium* symbiosis can provide a more economical source of N. The ability of legumes to fix nitrogen from the atmosphere, in association with the *Rhizobium* bacteria, gives the plants a dual-role of providing an improved diet to growing animals and of increasing soil fertility through release of nitrogen.

Legumes provide high-quality protein and energy, often critical during the dry season when animals, feeding solely on grass, lose much of the weight they gained during the wet season. Pasture scientists in the tropics had shown the potential and value of grass-legume pasture and an increasing number of cattle raisers are appreciating it.

CHARACTERISTICS OF TROPICAL PASTURES

Most of the native pastures in the tropics are dominated by grasses like *Imperata*, *Themeda*, *Chrysopogon* and *Aristida* which have a short growing period or mature very quickly compared to the temperate pasture species. Protein deficiency is very common in tropical grasses, particularly the native species (Humphreys, 1972). Crude protein of native pastures is usually below 3%, particularly during the dry season and, when crude protein levels fall below 7 to 8%, animal production would be limited by protein deficiency (Evans, 1968). In addition, most of the grasslands in the tropics are found in marginal areas with low fertility status and are invaded by non-palatable species which dominate the area with increasing grazing pressure.

The value of legumes in pasture

The nutritive value and digestibility of tropical legumes is higher than that of tropical grasses and the quality of herbage from grasses rapidly declines with increasing maturity. In contrast, herbage from the legumes remains good throughout the growing period, except for the fodder trees which become woody as they mature, although such a situation is easily overcome by regular lopping of the plants.

In addition most of the tropical legumes are more productive than the grasses during the dry season, making them more valuable as sources of additional high quality feed during this period, thus increasing the year-round carrying capacity of pasture.

Legumes in native pasture

A number of studies showing the value of tropical legumes to grazing animals has been done in Australia and recently several results have been reported from numerous grazing trials conducted in Thailand, Malaysia and the Philippines, with most of the studies done on liveweight gain of beef cattle. Most of the studies however are on beef production from legume-based pastures involving Centrosema (Centrosema pubescens), Townsville Stylo (Stylosanthese humilis), Siratro (Macroptilim atropurpureum), Schofield and Cook Stylo (S. guianensi), Seca Stylo (S. scatia),, Verpuo Stylo (S. hamata) and Ipil-ipil (Leucaena leucocephala).

Table 1 shows the summary of liveweight gain data obtained from various studies in the Philippines. The data indicated that animal production from native pastures can be increased two- to four-fold by incorporating suitable pasture legumes, particularly with the Stylos which thrive well in dry acidic conditions. In Mabate, Imperata pastures overseeded with Centro or Stylo can easily support 1 animal unit per hectare with more or less 100 kg liveweight gain per hectare per year, while pure Imperata pasture stocked at 0.5 and 1.0 animal unit per hectare produced only 22 to 25 kg LWG per year, coupled with the animal loss in weight during the dry season.

Table 2 shows beef production on improved grass/legume pastures in the Philippines. Napier/Centro pasture fertilized with 65-45-45 NPK produced 128-148 tons of fresh herbage per year in a study conducted at ANSA farms. This pasture safely carried three animals per hectare with beef production of 475 kg liveweight gain per hectare per year. With four animals, the pasture had 806 kg liveweight gain per hectare but ran out of grass for 37 days and corn-stover supplementation was carried out.

Cowan (1986) summarised the levels of milk production that have been obtained from tropical pastures in Table 3. The grass and legume mixed pastures gave higher production per cow than nitrogen fertilized grass pastures. However, they cannot carry as heavy stocking rates as nitrogen fertilized grass and production per hectare is lower. Average production per cow is of the order of 10 to 12 kg/day for Friesian cows, 7 to 9 kg/day for Jersey and 6 to 10 kg/day for crossbred cattle. Production per hectare varies from 2,600 to 8,300 kg/year. Grass and legume systems have produced up to 8,000 kg/ha/yr, but loss of legume was observed at high stocking rates (Cowan et al., 1975)

Table 1. Live weight gains on Imperata and Imperata/legume pasture.

Location/ animal	Pasture	Stocking rate (a.u./ha)	ADG (kg)	LWG/hd (kg)	LWG/ha (kg)
Masbate (cattle)	Imperata	0.5	0.12	43.2	21.6
	Imperata	1.0	0.07	26.6	26.6
	Imperata/Stylo	1.0	0.32	116.6	116.6
	Imperata/Centro	1.0	0.25	91.8	91.8
Bukidnon (cattle)	Imperata	1.0	0.21	77.4	77.4
	Imperata/Centro	1.0	0.26	94.1	94.1
Bohol (Carabao)	Imperata-Themeda	0.5	0.24	85.4	42.7
	Imperata-Themeda	1.0	0.22	78.9	78.9
	Imperata-Themeda/ Stylo	0.5	0.35	127.0	63.0
	Imperata-Themeda/ Stylo	1.0	0.25	92.2	92.2
Bohol (Carabao)	Imperata	0.75	0.22	68.1	51.1
	Imperata/Leucaena	1.5	0.35	111.9	167.9
	Imperata/Leucaena	2.0	0.28	87.4	174.8

Table 2. Liveweight gain production on improved grass/legume pastures.

Location	Pasture	Fertilizer rate (kg/ha/yr N-P-K)	Stocking rate (au/ha)	ADG kg	LWG/hd kg	LWG/ha kg
Bukidnon	Para grass /Centro	0-50-0	2.0	0.423	155.9	311.8
	Para grass /Centro	0-50-0	2.0	0.419	150.9	305.8
ANSA Farm (South Cotabato)	Napier/ Centro	65-45-45	2.0	0.428	156.5	313.0
	Napier/ Centro	65-45-45	3.0	0.431	158.0	474.0
Bohol	Guinea/ Cook Stylo	24-24-24	2.0	0.240	86.4	172.8
	Guinea/ Cook Stylo	24-24-24	2.5	0.280	100.8	252.0
	Guinea/ Cook Stylo	24-24-24	3.0	0.220	79.2	237.6

Table 3. A Summary of milk production per cow and per hectare from cows grazing tropical pastures.

	Breed	Stocking rate (cows/ha)	Milk yield (kg/cow/day)	Milk yield (kg/ha/yr)
Unimproved pastures	Jersey	1.1	6.8	2660
	Guernsey	1.5	6.9	2670
	Jersey/Criollo	1.0	6.9	2660
	Friesian/Zebu	1.5	6.9	2760
Improved grass-legume pastures	Jersey	1.0	8.5	-
	Guernsey	1.8	9.3	4700
	Friesian	1.6	12.4	5350
	Friesian/Zebu	1.7	7.3	3720
	A.F.S.	1.6	8.0	3840
Improved nitrogen fertilized pastures	Jersey	2.5	6.8	5250
	Guernsey	2.5	7.8	5350
	Friesian	2.5	11.0	8250
	Friesian/Zebu	2.2	8.7	5200
	Jersey/Criollo	2.6	6.7	4100
	A.F.S.	2.5	7.0	4800

TREE LEGUMES

Shrub and tree legumes are a good source of protein and have been gaining importance for livestock production in S.E. Asia and other tropical countries. Fodder tree legumes have several attributes that make their potential use in the tropics very high. The legumes have deep root system and can withstand drought and often serve as the main source of forage during the dry season. Some of the tree legumes are multiple purpose plants and are often grown for fuel wood, timber, poles and even a source of food, in addition to fodder. Tree legumes, once established, are easier to maintain in association with tropical grass compared to conventional creeping legumes and they can be grown as an upper story on land used for growing crops at lower levels. Some of the legumes with known use and potential as a source of fodder are species belonging to the genera of *Albizzia*, *Callandra*, *Gliricidia*, *Mimosa*, *Leucaena*, *Samanea* and *Acacia*. Among the legumes, *Leucaena leucocephela* is most well studied while the value of *Gliricidia maculata* is now being recognised, along with other species as a source of fodder.

In the Philippines, Ipil-ipil or Leucaena is the most popular legume and has been given a great deal of attention since the early seventies, with people in the livestock sector looking at it almost on a 'cure-all' for the growing animal industry. This bias worked against the sector because, in 1985, Leucaena was infested by a 'jumping lice' or psyllid (Heteropsylla cubana) and we were not prepared with an alternative species. Currently, the infestation is still around but not as destructive as in 1985-1987 when the intensive cattle-fattening, smallholder farms were badly affected, forcing them to reduce animal holdings or stopped operations.

The infestation also affected the feed milling industry which utilized Leucaena as a source of xanthophyll and carotene in mixed feeds. Likewise, it also affected the smallholder farmer who grew, harvested and sold the leaves to the feed merchants and feedmills.

Smallholder dairy farmers raising Sahiwal-Holstein Friesian feed their animals with 5 to 19 kg fresh Ipil-ipil leaves in combination with fresh grass fodder and obtain 4 to 7 kg milk per cow per day. Ipil-ipil is planted in hedges around the home-lots and farmlots, and in evenly spaced rows (1m to 2m) under coconuts.

Liyanage and Jayasundera (1988) reported that several trials conducted by the Coconut Research Institute in Sri-Lanka have demonstrated the value of Gliricidia as an animal feed. Gliricidia loppings mixed with Brachiaria milliformis in 50-50 ratio and fed to crossbred heifers resulted in an average live weight gain of 700 g/head/day. In another trial, a mixture of Gliricidia and Leucaena was planted alternately 1.5 m apart along the fence in a pasture/ cattle/coconut integrated system and produced more than 2 MT/ha/year of fresh green matter. This, when fed to heifers at the rate of 6 kg along with a pasture during the dry season, produced average live weigh gains of 300 g/head/day. Freshly chopped Gliricidia leaves can also reduce the duration of urea-treated straw from 21 to less than 6 days.

Gliricidia leaves are succulent but may not be very palatable to animals when first introduced. However, livestock freely eat when they become accustomed to the taste. Table 4 shows that Gliricidia, when fed with Siguaue (Brachiaria brizantha) grass from 0 to 100% for one month to Jersey milch cows, had no adverse effect on their health or milk production and was very palatable (Chadhokar and Lecamwasam, 1982). However, tainting of milk when Gliricidia is fed above 50% supplementation level has been reported but this may be avoided if feeding of this material is stopped a few hours before milking.

Table 4. Effect of Gliricidia maculata in a mixture with Brachiaria brizantha on milk yield and its composition

Treatment	Average Milk Yield (litres/cow/day)		Average Milk Composition (Percentage)		
Percentage Gliricidia	Pre-exptl.	Exptl.	Fat Pre-exptl.	Fat Exptl.	Solids not-fat
0	5.5	5.8	4.8	4.7	8.3
25	5.2	5.8	5.2	5.1	8.6
50	6.5	5.7	5.7	5.8	8.8
75	3.8	4.6	5.1	5.3	8.4
100	6.7	7.6	5.3	5.8	8.6

SUMMARY AND CONCLUSIONS

Vine and tree legumes are very valuable components in livestock feeding systems in the tropics. Increasing population trends indicate that more of the grassland areas will be diverted to crop production and will limit the use of pasture legumes to marginal areas where crop production is economically less or not feasible. Smallholder livestock production will increase in proportion and fodder trees will be socially and economically viable.

Livestock research and development programmes should focus on utilisation of tree legumes relevant to existing farming systems in smallholder farms in the tropics.

REFERENCES

Chadhokar, P.A. and Lecamwasam, A. Effect of feeding Gliricidia maculata to milking cows: A preliminary report. Tropical Grasslands 16 (1): 46-48.
1982

Cowan, R.T. Beef and Dairy production in the Wet tropics. 6th FAO Regional Training course. Pasture Management for the Wet Tropics of South East Asia Course Manual. pp. 193-197.
1986

Cowan, R.T. Byford, I.J.R. and Stobbs, T.H. Effect of stocking rate and energy supplementation on milk production from tropical grass legume pastures. Aust. J. Agric. Anim. Husb. 15: 740
1975

Evans, T.R. 1968 Source of nitrogen for beef production in the Wallum. Trop. Grasslands 2 (2): 192-115.

Humphreys L.R. 1972 The nutritive value of tropical pasture species and measurement of productivity. Lecture notes sets, AFFHC/FAO Tropical Pasture Science training course for South-West Asia. College of Agriculture, Malaya.

Liyanage, L.V.K. and Jayasundera, H.P.S. 1988 Gliricidia as a multipurpose tree for coconut plantations. Coconut Bull., Coconut Res. Inst., Sri Lanka Vol.1. pp. 1-4.

THE DEVELOPMENT OF DAIRY FARMING IN THAILAND

by

S. Pichet

BACKGROUND

Although dairy farming in Thailand began around 80 years ago, the introduction of extensive dairy development took place in the early 1960s. It started with the establishment of the Thai Danish Farm and Training Centre (TDDF) at Muak Lek, as a joint venture between the Thai and Danish Governments. The approach was the clearing of land, purchase of cattle, construction of farm buildings, training of farmers, development of a dairy colony, provision of extension services and development of a small dairy plant, as well as a marketing system for pasteurized milk production. In 1971, the Thai Government took over responsibility and the project was organised under the management of the newly established government enterprise, under the name of "The Dairy Farming Promotion Organisation of Thailand (D.P.O.)"

The objectives of D.P.O. are to promote milk production, to process milk and to sell milk products. Several important activities have been employed by D.P.O. to promote dairy farming. These include offering crossbred heifers at cost price to newly established dairy farmers, training of people wanting to become dairy farmers, provision of extension services including artificial insemination, veterinary services, milk recording, farm management advice, a milk collection centre and the buying of milk at guaranteed prices.

Milk production in the Nong Pho area started almost at the same time as in Muak Lek, but in a different way. The farmers were already established there, with smaller pieces of land, but they were more progressive and had the assistance of the Department of Livestock Development (D.L.D.). Crossbreeding was successfully employed through the use of artificial insemination. In 1971, a cooperative dairy plant was built and the milk production, the organisation of milk processing and milk marketing have all been most impressive.

The Thai-German Dairy Training and Processing Plant in Chiang Mai was established in 1968 and operated as a joint venture until 1977, at which time it was taken over by the Department of Livestock Development (D.L.D.). Since 1979, D.P.O. has been responsible for the processing and marketing functions.

In 1982, D.P.O. started the dairy farming project in the South, which included construction of a dairy plant for UHT milk at Pranburi, Prachuab Kirikan. The purpose of the project was to reduce the area used for growing pineapples, because farmers were earning inadequate incomes due to over-supply of the product. In addition, waste from the pineapple canneries could be used as a cheap roughage for dairy cattle.

GOVERNMENT POLICIES FOR DAIRY FARMING

Current dairy policies approved by the Agricultural Policy and Planning Committee on 20th February, 1987 are as follows:
- to promote the expansion of raw milk production in order to reduce importation of dairy products,
- to increase the efficiency and quality of dairy farming, milk collection, milk processing and milk marketing,
- to strengthen dairy cooperatives as profit oriented organisations,
- to encourage the private sector to be involved in dairy development, and
- to organise an independent Milk Board with representation from all sectors of the dairy industry which would execute policy on an industry-wide basis.

The Government's plan for the development of dairying is aimed at a reduction of foreign exchange for the purchase of imported dairy products but also to provide the farmers with the opportunity to earn increased and more regular incomes and generate employment opportunities in farming, milk processing and manufacturing industries. The target is to produce 328,000 tons of raw milk by 1996, which can meet half the demand for dairy products. To achieve this, it will, for instance, be necessary to increase the number of dairy cows to 117,000, which will require a growth rate of 18% per annum (Table 1). The goal could be reached in two ways: by increasing the local supply of dairy cows through the insemination of local Brahman-cross cows and by purchasing from overseas.

THE GROWTH OF THE RAW MILK PRODUCTION FROM 1982 TO 1987

Over the past 25 years, the Government's support to dairy farming through D.P.O. and other key government departments has been successful. For instance, in the period from 1982 to 1987, the production of raw milk has increased annually by almost 24% to 79,000 tons by the end of the period. The number of dairy cattle has, over the same period, increased by approximately 20% annually to 75,500 head in 1987 (Table 2).

Ever since its establishment, the D.P.O. has, together with the Nonh Pho dairy cooperative, dominated the production and utilisation of raw milk. In 1985, D.P.O. had a share of 66% of the raw milk production, 66.8% of the dairy cattle population and 58.4% of the dairy farmers are suppliers.

Table 1. Plan target for raw milk production 1988-1996

Year	Number of cows	Raw milk Produced
1988	40,840	91,000
1989	46,680	109,000
1990	51,880	125,000
1991	59,850	147,000
1992	68,800	174,000
1993	78,660	204,000
1994	89,840	240,000
1995	102,900	281,000
1996	117,130	328,000
Average growth rate Per Annum (%)	14	18

Source: Office of Agricultural economics

Table 2. Dairy cattle numbers and raw milk production in 1982-1987

Year	Total Female cattle (head)	Calves and heifers (head)	Dairy cows (head)	Milk Production tons/yr	tons/day
1982	30,046	16,280	13,766	27,027	74
1983	39,426	20,233	19,193	36,030	99
1984	48,489	24,639	23,850	46,197	127
1985	50,988	26,410	24,578	54,560	149
1986	69,907	36,479	33,428	69,175	190
1987	75,500	39,300	36,200	79,100	210
Average increase per year	19.87	19.82	22.22	23.87	23.87

Source : Office of Agricultural Economics

DEMAND FOR RAW MILK PRODUCTION

The scope for development of the dairy industry is excellent, because the production of raw milk in 1989 is still only a fraction of the total demand. If all milk products consumed in 1986 had been made in Thailand, at least 500,000 tons of raw milk would have been required. The actual production of raw milk was 69,000 tons or 14% of the demand.

The raw milk collected by D.P.O. and the cooperatives is processed into ready to drink (RTD) milk or sold in bulk to private dairy companies, and the latter also use the raw milk for RTD milk, which includes pasteurized milk, U.H.T. milk, canned sterilized milk, etc. In the first instance, the raw milk production could be used to meet the demand for RTD milk.

Table 3 shows how the position would be then: the production of raw milk in 1986 was 69,200 tons (14%) of the total demand, while RTD milk consumption was 81,600 tons or about 17% of the total demand. It is interesting to note, that the consumption of RTD milk went up to 126,300 tons in 1987. This increase is the result of the milk consumption promotion campaign supported by the Government and the private dairy companies.

Table 3. Raw Milk production and consumption of Ready-to-Drink Milk in 1982-1987.

Year	Raw Milk Production '000 tons	RTD milk Consumption '000 tons	Deficit of Raw milk '000 tons
1982	27.0	44.4	17.4
1983	36.0	58.4	22.4
1984	46.2	62.4	16.4
1985	54.6	66.0	11.4
1986	69.2	81.6	12.4
1987	79.1	126.3	47.4

CONSTRAINTS ON RAW MILK PRODUCTION

One constraint on the raw milk production is the high costs of some inputs required in the production, in particular concentrate feed. The percentage contribution to the total variable costs in the period January - June 1986 were: concentrates 60%, roughage 16%, disease 2%, labour 15% and others 7% (Source: Office of Agricultural Economics).

Another constraint is shortage of dairy cattle: the growth in the number of local crossbreds is insufficient and imported cattle are rather expensive. The price (CIF) of imported New Zealand crossbred heifers (5-7 months pregnant), in 1987 and 1988, was 860 and 910 US dollars respectively. Early in 1989, the cost climbed further to 960 US dollars.

Ways to increase the productivity and reduce costs of production would be to:

- improve feeding management. This would involve more use of improved grass/legume pastures, which could result in a reduction of the need for concentrates, as well as reduce the level of nitrogen fertiliser use. Furthermore, the pasture conservation could be improved and the use of by-products from crop production and/or agro-industries could be increased.

- improve management of the animals to improve the conception rate and reduce health problems, in particular mastitis.

- improve farm advisory services. The farm advisers should be familiar with progress in research and technology and should advise the farmers on the application of new techniques.

- improved breeding programmes. For instance, a national progeny test scheme for locally selected bulls should be developed and thereby reduce the need for expensive overseas semen. Implementation of better breeding programmes would also result in higher milk yields.

- improve systems for handling excess stock (bull calves, excess female calves and cull cows). Bull calves could be raised for dairy beef production.

COSTS OF RAW MILK PRODUCTION

Much greater emphasis must be placed on developing the required knowledge and skills to introduce and manage high quality improved pastures successfully. Farmers should understand and appreciate that there is a great deal of basic scientific evidence from many overseas countries showing that animal performance, and especially milk production, is much more dependent on the quantity and quality of feed eaten rather than on the genetic make-up of the animal. It is only when the level of feeding is high, in both quantity and quality, that the value and importance of good breeding generally becomes apparent. In other words, while dairy farmers should certainly give some attention to their breeding programme and the selection of semen, much more attention should be given to developing and improving their feed quantity and quality, particularly of their pastures which are the cheapest and most common source of feed for dairy cows in Thailand.

This view is supported by data obtained from a recent experiment at D.P.O. (Muak Lek). The cows involved were all 62.5 to 75% Holstein-Friesian or Red Dane crossbreeds in their 1st, 2nd or 3rd lactation and fed different levels of concentrate according to milk production, ranging from improved pasture only (no concentrate) to _ad libitum_ concentrate feeding, with 10 cows on each level of feeding. Detailed measurements of milk production, milk fat, animal health and pasture utilisation were made during both early and late lactation. Prior to the experiment a total of 50 cows were balanced across the five treatments, in order to ensure that there was no bias in production which might unfairly favour one treatment over another.

In this example, three comparisons were made: of cows fed improved pasture only; of cows fed 1 kg of concentrate to 3 kg of milk (3:1) plus improved pasture; and of cows fed _ad lib_. concentrate plus improved pasture. It is assumed that the cows ate approximately 4 kg DM per 100 kg body weight daily, which meant that the average 350 kg liveweight cows each ate approximately 14 kg of pasture DM on the pasture only treatment, approximately 10-11 kg of pasture DM plus 3-4 kg of concentrate in the 3:1 treatment and approximately 13-15 kg of concentrate plus negligible pasture in the _ad lib_. concentrate treatment.

These data plus the more obvious economic parameters are presented in Table 4. It is clear that dairy farmers can achieve a highly economic return from cows fed pasture only, provided that the pasture is properly managed, leafy and hence of high quality. A production of 2,400 kg of milk can be obtained from just "average" cows and up to 3,150 kg from "good" cows, yielding an estimated "profit" of 14,400 Baht and 20,420 Baht respectively per lactation. When a relatively small input of concentrate was also fed along with improved pasture, milk production was increased and more importantly, profitability was also increased. When cows were fed to appetite on meal concentrate of 15.5% crude protein and an estimated TDN of 70%, milk production was further increased but profitability was decreased. Obviously with the present price of meal concentrate farmers will achieve greater profit by relying on improved pasture as their main source of feed for dairy cows plus a small input (3:1) of concentrate, rather than striving for higher milk production from only concentrate but at higher cost.

Table 4. Estimated returns and costs of production from "average" and "good" cows on different feeding regimes.

	"Average" Cows	"Good" Cows
Improved pasture only:		
Milk yield/cow/d kg.[1]	8.0	10.5
Total lactation kg.	2400	3150
Milk sales (Baht)[2]	16080	21105
Cost of grass (Baht)[3]	1440	1440
Profit per cow (Baht)	14400	19605
3:1 concentrate + improved pasture:		
Milk yield/cow/d kg.[1]	10	13
Total lactation kg.	3000	3900
Milk sales (Baht)[2]	20100	26130
Cost of concentrate (Baht)[4]	3300	4290
Cost of grass (Baht)[5]	1220	1220
Profit per cow (Baht)	15780	20620
Ad lib. concentrate:		
Milk yield/cow/d kg.[1]	13	17
Total lactation kg.	3900	5100
Milk sales (Baht)[6]	25740	33000
Cost of concentrate (Baht)[4]	12870	14850
Profit per cow (Baht)	12870	18150

[1]300 day lactation
[2]6.70 Bht/kg (6.50 + 0.2 for fat%)
[3]14 kg DM/day, 0.4 Bht/kg
[4]3.30 Bht/kg
[5]10.5 kg DM/day, 0.4 Bht/kg
[6]6.60 Bht/kg (6.5 + 0.1 for fat%)

CONCLUSIONS

The efforts of the Government in the development of dairy farming in Thailand have been successful. There has been a satisfactory increase in milk production and in the number of dairy cattle and dairy farmers, so that today Thai farmers produce 15% of the raw milk needs for all milk products. By 1996, they could be producing 50%. Such a rapid growth requires great investments in facilities for the farmers. For instance, it will be necessary to import dairy cattle to get enough animals and to get the right stock. Furthermore, training of farmers and good extension services are important factors, especially in making dairy farming into a good business. Serious efforts must be made to reduce the costs of milk production and, as concentrates are expensive, it would pay to look for better utilisation of pastures and fodder crops.

MILK PRODUCTION SYSTEMS BASED ON PASTURES IN THE TROPICS

by

Roberto García Trujillo

THE LATIN AMERICAN REGION

The humid and semi-humid tropical areas of Latin America comprise approximately 70% of the total area of the region and are where 60% of the cattle are concentrated. Four different sub-areas exist, these are: a) native savannas, b) the "cerrados", c) tropical humid and semi-humid forests with fertile soils and d) tropical humid and semi-humid forests with non-fertile soils.

80% of the soils in these zones are barren, especially those of the savannas and "cerrados" where improved pasture establishment requires high levels of fertilizers and soil improvement. Rainfall is a limiting factor for pasture development, due to the unequal distribution throughout the year and the fact that the dry season can last for almost seven months, except in the humid forest area. There are dairy cattle in the region (3-11 cows/100 inhabitants) but, because of the low productivity (750-1700 litres/cow/year), large amounts of milk and milk products are imported.

The present situation in the foreign market, the general economic crisis and different aspects of agricultural policy in the countries of the region are against the development of milk production and cattle production in general. These problems were exhaustively analyzed in the First Meeting of Cattle Development in Latin America and the Caribbean, sponsored by FAO in September 1988 in Montevideo. There are many different milk production systems in the region, but the non-intensive pasture systems with low milk producing cattle predominate, while the most productive and the most intensive systems are mainly found in the high tropical areas or in production systems using mainly European breeds, where cattle remain housed or semi-housed and are fed forages, hay or silages and a high proportion of concentrates, supplemented with imported raw materials.

In Cuba, milk production is based on the utilization of pastures in the rainy season and green and preserved forages and sugar industry by-products in the dry season. Holstein dairy cows and their crossbreds with Zebu cattle are employed.

Although research on pasture production, feeds, production systems and other aspects is scarce, results are available which show that it is possible to considerably improve milk production in the tropics. The objective of this paper is to offer the results of some studies carried out in Cuba.

MILK PRODUCTION SYSTEMS BASED ON PASTURES

In the humid tropics, where rainfall is not limiting and declines to critical levels for only 2 to 4 months, milk production based on pastures is limited by the quality of pasture and the direct or indirect effects of the climate on the more productive animals. In the semi-humid tropics, the lack of rainfall during the dry period (lasting from 5 to 7 months), seriously limits pasture production and quality. Under these conditions, which prevail in the majority of the cattle breeding regions in the tropics, the development of a more productive milk production system should consider the equilibrium between the needs of the animals and feed production. The basic elements are pasture management, fertilizer use, stocking rate, forage production and supplementation, all being related to the animal potential.

Natural grasslands and their productivity

Natural or semi-natural, non-fertilized grasslands in areas with soils of low fertility allow a low stocking rate which, in general, does not exceed 1 cow/ha and is frequently about 0.5 to 0.8 cattle units per ha. Milk production recorded in the sub-humid area does not exceed 4 kg milk/cow/day and 300 kg of milk ha/year when Zebu cattle are used, as in the case of the western plains of Venezuela (Capriles, 1982), or from 5 to 7 kg milk cow/day and 1600 kg milk/ha/year when crossbreds are used and the animals are fed medium to low quality forage in the dry season, as reported by Blydestein _et al_. (1969) in Costa Rica or in certain dairy areas of the eastern provinces of Cuba (Table 1). In general, these herds are characterized by short lactations, poor calving rates and low percentages of cows in milk.

The low milk production of the animals in the grazing system with natural or semi-natural pastures is not only a consequence of this, but of the poor resources, operations and techniques with which these animals are managed and where supplementary feeds play an important role in their survival on account of the poor pasture quality or amount.

Under these conditions, legumes could be one of the most economic ways to improve the production level of these herds. In this sense, Monzote _et al_., (1985), with the introduction of _Glycine_ into natural pasture in areas of low rainfall (800 mm/year), found an increase in milk production from 1041 to 1684 kg/lactation, compared to cows on natural pasture. In the rainy season, the association produced from 3 to 5 kg of milk/cow/day more than with natural pasture although, during the severe months of the dry season, this difference was only 0.3 to 1.2 kg of milk/cow/day.

Another alternative under these conditions could be the use of forage areas of sugar cane as a supplement during the dry season.

Table 1. Milk production results under different grazing systems in the tropics.

Pastures	Area	Irrigation	Fertilization kg/ha/year	Supplementation		Breed	Stocking rate	Milk production kg	
				Forage	Concentrate			Lactation	ha/year
Bative	SH savannah	none	none	low	none	Zebu	0.5-0.8	240-350	120-300
NP+legume association	SH	none	P and K	low	none	xbred	1	1650	990
Native	SH	none	poor	moderate	low	xbred	1-1.8	1300	1600
Semi-natural	H	none	none	none	low	xbred	1	1800	1600
Improved	SH	none	moderate	high	low	xbred	2.5-3.3	1700-2500	5300-7300
Improved	SH	yes	high	none	none-low	xbred	2.7-4.5	2000-2400	6000-9000
Improved	SH	yes	high	low	none-moderate	Holstein	2.0-4.0	3000-4200	8500-14000
Improved	H	yes	high	low	low	xbred	4.0-6.5	2300-2700	7600-14000
Improved + legume	SH	none	low-medium	medium	none-low	xbred	1-1.5	2100-3200	2700-3800
Improved + legume	SH	yes	low-medium	none	none-low	European	1-1.5	3300-4200	4500-6000
Improved + tree legume	SH	none	medium	high	low	xbred	2.5-2.7	2500-3200	6000-9600
Improved + tree legume	SH	yes	high	low	medium	Holstein	2.5-3.0	3900-4300	7000-10500
Improved	SH	yes	high	medium	high	Holstein	4-5	4400-4600	17000

SH = semi-humid H = humid Fertilization: low 50-80, medium 130-180, high 240-350
Concentrate kg/cow/day low: less than 2 medium: 2-3 high: more than 3

Improved pastures

The utilization of improved pastures and fertilizers markedly increases stocking rate capacity, milk production per ha and individual yield, especially when breeds and crossbreds of medium to high milk potential are used.

In general, stocking rate is increased to 2.2 - 4.5 cows/ha in the sub-humid regions according to the type of pasture and the level of fertilizer. Milk production/cow is increased to 6 - 8.5 kg/cow/day and up to 2500 kg/lactation with crossbred animals and from 10 to 14 kg of milk/cow/day (3000-4000 kg/lactation) with European breeds. Milk production/ha could reach or exceed 12000 kg/ha/year (see Table 1).

In the humid regions, the stocking rate of the grassland could reach 6.5 cows/ha with 14000 kg of milk/ha/year and 8.5 to 9 kg/cow/day, as demonstrated in some areas of the region (Cubillos et al. 1975, Muñoz et al. 1988b).

The effectiveness of the systems based on improved pastures depends on the equilibrium between pasture selection, the level of fertilizer use and stocking rate. An imbalance in this sense could ruin the system and the stocking rate is a decisive factor in the productivity of dairy herds. Different trials have demonstrated that, when medium potential animals are used, the production per hectare can be employed as an indicator of productivity but when high potential cows are used and a higher production per hectare is obtained, the potential of the animals is only exploited to the extent of 50 to 60% (Figure 1), provoking a negative effect on reproduction and animal health.

The level of stocking rate and N fertilization are closely related. Our results show that from 50 to 100 kg N/ha/year, according to the stocking rate and system employed, are needed for soils of medium to low fertility per cow. A reduction in stocking rate, under Cuban conditions, has markedly improved individual milk production, without affecting milk production per ha, fertility, survival and health (Table 2).

The most productive pastures found are coastal bermuda grass with low stocking rates (2 to 2.5 cows/ha), common guinea grass and likoni for medium stocking rates and star grass with high stocking rates (4 to 5 cows/ha). Coastal bermuda grass requires a high level of fertilizer, irrigation in the dry season and careful management. Also, short grazing rotations in the rainy season (12 to 18 days), together with night grazing, allow ample selection, availability of

Figure 1. The effect of stocking rate and the type of pasture on milk production per cow and per hectare (from Pérez Infante, 1971 and others in Cuba).

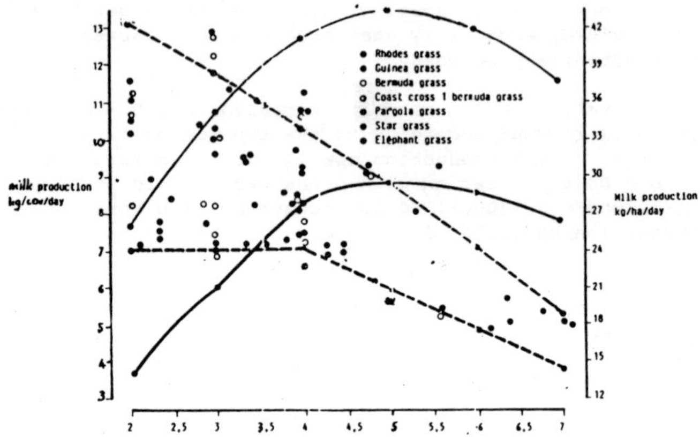

Table 2. Stocking rate (SR) reduction and performance of various dairy herds in Havana Province (Martinez, 1980; Salinas, 1988, unpublished).

	SR cows/ha	Improved pasture %	Milk yield (kg) /cow	/hectare	Fodder supplement DM/cow/d	Animal losses %	Birth rate %
Case A: Three dairy units							
Unit A	2.84	33	1082	3078	4.2	3	58
Unit B[1]	2.26	38	1650	3741	3.8	6	67
Unit C	2.24	73	2102	4624	2.2	4	76
Case B: Dairy enterprise (15000 cows) then reduce stocking rate							
Previous	2.9	-	2428	7043	-	11	68
After	2.5	-	2662	7037	-	9	72
Case C: Dairy district (3800 cows) then reduce stocking rate							
Previous	3.3	-	2203	7279	6.4	13.8	-
After	2.2	-	2844	7580	4.4	10.0	-

[1]Stocking rate reduced one year previously.

pastures (50 kg DM/cow/day) and good milk production in European breeds. In the dry season, overgrazing should be avoided by restricting grazing time (3 to 4 hours/day) or housing some of the animals.

FORAGE PRODUCTION

In the sub-humid tropics, large amounts of forage are used daily to make up for pasture deficiency in the dry season. In Cuba, various alternatives have been studied and used to this end. They can be classified as follows:

I Sowing areas of forage with irrigation so as to produce silage in the rainy season and green forage in the dry season.
II Conserving pasture surplus in the rainy season to produce silage and hay supplies for the dry season.
III Sowing forage areas with sugar cane which is harvested 12 months later during the dry season.
IV Utilization of coarse by-products of the sugar industry (bagasse -molasses-urea, pre-digested bagasse).

The main forage used in Cuba is king grass, which under production conditions, yields from 97 to 130 t/ha/year when fertilizer (300 to 350 N kg/ha) and irrigation in the dry period are used. The application of manure (25 to 30 t/ha/year) favours forage stability, decreases the need for N fertilizer (100 to 150 kg H/ha/year) and the use of P and K. Because this forage has a low DM content (18%) when it has a good nutritive value (from 10 to 11% CP), it is necessary to pre-wilt it so as to obtain a good quality silage.

Star grass and guinea grass are also used as the main forage with good yields and supplementary forages such as sorghum and sunflower are used as pure crops or are intercropped with pastures, new sowings or forage areas during the dry period.

In general, green forage is not used as the sole source of roughage in the dry season, except for some high producing herds or high producing groups of cows and when there is great availability, being employed as a supplement to silage diets and sugar cane (30% green forage: 70 % sugar cane or silage) (Esperance and Perdomo, 1978, Muñoz et al., 1988a).

Medium quality silages, which are more abundant in commercial units, when supplied together with grazing on non-irrigated pastures, produce from 3 to 5 kg of milk/cow/day, but the animals lose weight and have short lactations. In order to obtain an average of 8 kg of milk, supplementation of 1.2 - 2 kg of concentrate is required with the addition of 30 - 40% of total roughage as green forage. Above this level of supplementation, a response of 1.5 kg of milk/kg of concentrate could be expected (Gutierrez et al., 1988) (Figure 2).

Figure 2. Milk production from tropical silage or forage of medium quality (adapted from Gutierrez et al., 1988).

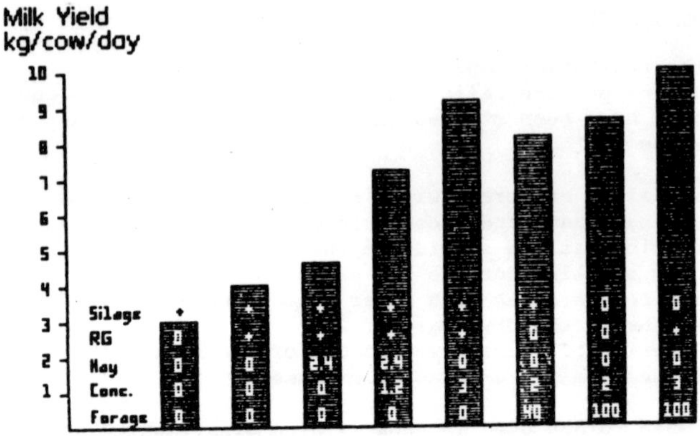

Milk production under non-irrigated conditions with animals fed silages is strongly related to the DM percentage of the silage and its consumption, an increase of 0.3 kg of milk and 0.37 kg of DM consumed per percentage unit increase in dry matter being reported.

Another difficulty with tropical silages is the loss of nutrients and materials produced during silage processing, when silages are not made with adequate techniques. Low capacity silos with long cut forage and high humidity could lose almost 60% of the material. However, when high DM and finely cut forages are used and the material is placed in adequate places or bunker silos, the losses could be diminished to 15% at the most.

Sugar cane is characterized by its high yields (up to 170 t fresh matter/ha with low N fertilization (100 to 150 kg N/ha/year) and is harvested during the dry period, thus there are no conservation expenses. Although used as forage, low consumptions are observed, particularly when it is not adequately mixed with other forage feeds and supplemented with NPN and minerals. When using sugar cane for animal diets, the material should be finely cut and 1% urea (Perez Infante and Garcia Vila, 1975), 0.12 to 0.15% sulphur (Ruiz, 1979) should be added and also it should be combined with 30% forages, pastures or both (Muñoz et al., 1988a), and the necessary concentrate supplementation to cover the animal requirements.

Experiments carried out in the Caribbean area (Garcia and Neckles, 1983; Perez Infante, 1975; Garcia Trujillo et al., 1981 unpublished; Senra et al., 1988 unpublished) show that as the percentage of sugar cane in the DM of the ration is increased, milk production decreases by approximately 1.46 kg milk for each 10% of sugar cane inclusion, yielding from 7 to 10 kg of milk/cow/day when 50% of sugar cane is added. This indicates that feeding factors and the animal potential could vary the response (Figure 3).

Figure 3. Sugar cane forage level in the diet and milk yield.

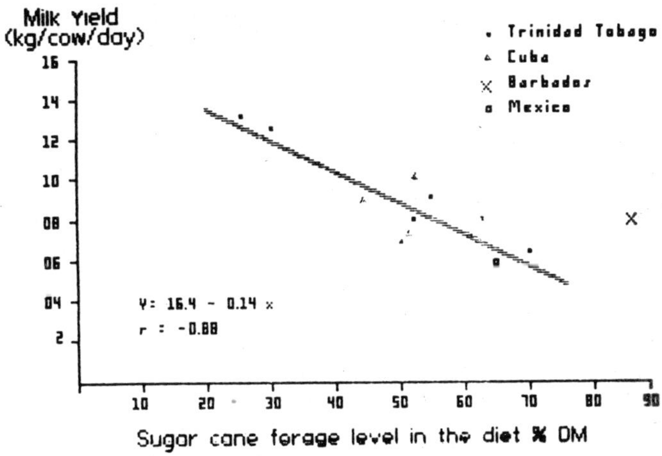

The sugar cane industry produces various fibrous by-products for animal feeding. The most widely used in Cuba is bagasse, pre-digested with 2 - 3% sodium hydroxide, to which 15% molasses and 10% urea are added, and which finally contains 8.37 MJ ME/kg DM and from 11 to 12% CP. This feed is used as forage in the dry season and consumed at a rate of 6-7 kg/cow/day.

At Indio Hatuey Branch Station, a comparison between the forage systems I, II and III (see above), crossbred (3/4 Holstein, 1/4 Zebu) cows on pasture without irrigation for 3 years showed higher milk production per lactation than on the sugar cane system (2565 vs 2427 kg), but the sugar cane system produced higher fertility (80 vs 67-72% cows pregnant per year), lower cullings (11 vs 22-25%) and thus the milk production/ha was higher (7704 vs 6640 kg) (Table 3).

In a later study where systems II and III included a protein supplement and stocking rate decreased to 2.5 cows/ha, milk yield per lactation was increased by 150 and 424 kg for the conservation and sugar cane systems, respectively. Protein supplementation resulted in calving intervals of 393 days on the sugar cane system and reduced both the use of concentrates and the cost of production in 10%.

The comparison of system I, III and IV with Holstein cows at ICA showed a higher milk production per cow on the silage system than on the sugar cane or bagasse pith systems (3553 vs 3100 kg/lactation and birth rate (87 vs 83%), but the production cost was slightly higher (0.22 vs 0.20 Cuban pesos/kg of milk).

SUPPLEMENTATION

Commercial concentrate supplementation is practised in dairy units using medium to high potential cows. The response to concentrates obtained, in terms of milk production, ranges from 0.2 to

Table 3. Comparison of three milk production systems (García Trujillo, 1981, unpublished).

	With conservation areas	King Grass forage areas	Sugar Cane forage areas
Milk production			
kg/lactation	2537	2594	2427
kg/cow/year	2219	2209	2348
kg/hectare	6657	6727	7044
Fertility			
cows pregnant/year %	67	72	80
Culled cows %			
reproductive problems	14	22	11
total	22	25	11
Feeding cost			
(pesos/kg milk)	0.088	0.084	0.083

2.3 kg of milk/kg concentrate. This response mainly depends on the differential between the production potential of the basic diet and the cow's potential (Garcia Trujillo, 1988). Our experience recommends the use of limited amounts of the available concentrates 21 days before calving and in cows producing more than 12 kg of milk per day, although if basic diets of poor quality are used, economic responses to supplementation can be obtained but foreign currency will be wasted.

Legumes are the main crops available for total or partial substitution of concentrates in diets for dairy cows. Pereiro (1985) demonstrated, in a series of experiments, that with 30-38% of the total area in Glycine wightii with fertilized and irrigated coastal bermuda grass, irrigated in the dry season and grazed daily or on alternate days (3 hr daily), with 3.8 kg concentrate/cow/day, production of 13 kg milk with 2.8 cow/ha was achieved. Milk productions from 12 to 14 kg/cow/day were also obtained when Glycine was used as a protein supplement given to cows consuming diets of forage, silage and hay.

In the Institute of Animal Science in Cuba, Leucaena and Glycine have been used with 3.2 and 2.7 cows/ha respectively, with a 50% reduction in the concentrates needed to produce approximately 14 kg of milk/cow/day. Furthermore, good quality milk, a high calving rate (86%) and no health problems were observed in these cows.

Other alternative supplements have been developed using sugar cane by-products and sugar cane (Table 4). Muñoz (1982) worked with nitrogenous activator supplements (NAS) formed by filter cake mud (45%), molasses (40%), urea (11%) and minerals (4%), supplied to a dairy herd consuming fertilized pasture in the rainy season and non-irrigated pasture plus silage in the dry season, and produced 10.3 kg milk/cow/day with a calving rate of 83%. NAS increased the digestibility and consumption of low and medium quality basic diets.

Table 4. Some supplements produced from by-products of sugar cane industries or sugar cane.

Products	Formulas %					
	1	2	3	4	5(NAS)	6
Saccharina	90	70	-	-	-	-
Sun-dried sugar cane	-	-	50	-	-	-
Dried Filter Mud	-	-	-	30	40	-
Molasses	5-6	5-6	15	15	35	69
Cereals	-	1C	20	41	-	21
Proteins	-	10	10	10	0-10	4
Urea	0-1	0-1	1	1	10	3
Minerals	4	3	4	3	5	3

NAS: Nitrogen activated supplement   Supply: 350 g/kg milk
Except NAS (1.5-2.5 kg cow)

Sugar cane dehydrated in the sun (50%) or dry filter cake mud (30%) have been used at lCA to produce supplements where the remaining components were molasses, cereals and minerals. These supplements are supplied at a rate of 300 g per kg of milk and allowed the production of 9 to 10 kg of milk during the dry season. More recently, a new product obtained from clean sugar cane, without leaves, enriched with protein, has been obtained by an aerobic fermentation process. This new product was developed by Dr. Arabel Elias at our Institute of Animal Science and has been named "Saccharina". Saccharina can have between 9 to 11% of CP and approximately 10.4 MJ ME/kg DM for ruminants. It is used to supplement all animal species and from 10 to 12 kg of milk per cow per day have been obtained in dairy cows supplemented with this product.

CONCLUSIONS

In the system based on pastures, stocking rate adjustment is of vital importance so as to have a stable and productive system. 90% of the animal potential should be obtained from pastures and supplements. Stocking rates should also be adjusted to avoid surplus forage supply during the rainy period and the surplus conserved for the other months. Pasture availability of 50 kg DM/cow/day should allow the exploitation of the most productive pastures.

In the semi-humid areas of the tropics, the production of forages to cover pasture deficiency in the dry season is necessary. There are various options and one of them is to use sugar cane for crossbred cattle in small units. In large units or with cattle with a higher potential, silages and green forages should give better results, if these are of good quality.

Legumes are essential components of any system of dairying. The grass-legume pastures are better with low stocking rates, while with medium and highly stocking rates, forage legumes are preferred. Legumes save concentrates and fertilizers and improve the productive performance and health of the herd.

Commercial concentrates are very expensive thus these must only be used in those animals with the required physiological status (European breeds, end of the pregnancy, beginning of lactation).

Low producing crossbred cattle (10 kg milk/cow/day) can be fed with by-products obtained from sugar cane.

REFERENCES

Blydestein, J., Louis, S., Toledo, J. and Camargo, A. Productivity of
1968 tropical pastures. 1. Pangola grass. Journal of the British Grassland Society 24: 71-75.

Capriles, M. Sistemas de producción de leche y carne para los llanos
1982 occidentales de Venezuela. lnforme Final IPA. Universidad Central de Venezuela. pp. 141-176.

Cubillos, S., Muñoz, H., Ruiz, M., Deaton, O. and Fuentes, G. Un
1979 sistema de producción de leche para pequeños productores. Resúmenes VII Reunión Asociación Latinoamericana de Producción Animal, Panamá, pp. 8-11.

Esperance, H. and Perdomo, A. Ensilaje y/o forraje para la producción
1978 de leche. Pastos y Forrajes 1: 415-436.

FAO Examen del sector ganadero de la región de America Latina y el
1988 Caribe. Informe de la Comision de Desarrollo Ganadero para America Latina y el Caribe. Primera Reunion. Montevideo, Uruguay. 20-30 Sept. 1988.

García, G. and Neckles, F.A. Feeding sugar cane for the productions of
1983 meat and milk. Feeding of Animals in the Caribbean. A workshop CARDI and SPC. Trinidad WI, pp. 88-100.

García-Trujillo, R. Potencial y utilización de los pastos tropicales
1983 para la producción de leche. In. Los Pastos en Cuba. Tomo 2. Utilización. J. Ugarte, R.S. Herrera R. Ruiz, R. Garcia, A.M. Vázquez and A. Senr. (Eds.). EDICA pp. 247-298.

García-Trujillo, R. A study of concentrate supplementation to dairy
1988 cows. Cuban Journal of Agricultural Science 22: 38-46.

Gutiérrez, A., Muñoz, 8., Esperance, M. and Michelena, J. Utilización
1983 del ensilaje en la producción animal. In. Los Fastos en Cuba. Tomo 2. Utilización. J. Ugarte, R.S. Herrera, R. Ruiz, R. Garcia, C.M. Vázquez and A. Senra (Eds.). EDICA pp. 437-476.

Monzote, M., Aira, A., Gómez, I., Rill, S. and Barbón, J.
1985 Comportamiento de la asociación pasto natural/glycine a nivel de extensión. Asociación Latnoamericana de Froducción Animal. Memoria. Volumen 20. p. 81.

Muñoz, E. Producción de leche y digestibilidad en vacas alimentadas
1982 con gramineas tropicales suplementadas con altos niveles de nitrógeno no proteico en secano. Thesis Candidate to Doctor in Veterinary Science. Agricultural Science Insti- tute for Higher Education, La Habana, Cuba.

Muñoz, E., González, R., Ruiz, R. and Galindo, Juana. Uso de la caña
1988a de azúcar para la producción de leche. Informe final de etapa de investigación. Instituto de Ciencia Animal. La Habana, Cuba.

Muñoz, H., Ramsammy, P. and Lallapachan, V. Searching for appropriate
1988b technology in dairy production. I. St. Stanislaus dairy unit milk production performance. Sugar Cane Feed Center Workshop. Trinidad and Tobago, Sept. 11-15.

Pereiro, A. Utilización del pastoreo restringido de glycine
1985 (Neonotonia wightii) como suplemento a vacas lecheras de mediano potencial suplementadas a base de pastos o forrajes conservados. Thesis Candidate to Doctor in Veterinary Science. Agricultural Science Institute for Higher Education, La Habana, Cuba.

Pérez-Infante, F. Efecto de diferentes especies de pastos y sus
1971 combinaciones en la producción de leche. Memorias Microestación de Pastos Niña Bonita. INRA, La Habana.

Pérez-Infante, F. and Garcia Vila, R. Uso de la caña de azúcar en la
1975 alimentación del ganado en época de seca. 1. Efecto de la adición de urea en el consumo y producción de vacas lactantes. Revista cubana de Ciencia Agricola 9: 109-112.

Ruiz, E. Utilización de la caña de azúcar en la producción de leche.
1979 Thesis Candidate to Doctor in Veterinary Science. Agricultural Science Institute for Higher Education. La Habana, Cuba.

DAIRY PRODUCTION IN THE SEMI-ARID RANGELANDS OF WEST AFRICA

by

Modibo Traoré

INTRODUCTION

The semi-arid zone of West Africa covers nearly 1.5 million km^2 between the arid Sahara desert in the north and the Sudanese savannah in the south. It includes a small fringe of the Sahel and a larger Sahelo-Sudanese area in the south. There are two main seasons a year: a short (3-5 mo.) rainy season starting in June and a longer (7-9 mo.) dry season. Annual rainfall varies from 500 mm in the north to 900-1000 mm in the Sahelo-Sudanese south.

THE ENVIRONMENT

The soils are generally sandy and of windblown origin in the north. Their organic matter content is low. On the rocky outcrops and on the laterite crusts, one finds shallow skeleton soils, rich in major minerals. The dunes, the rocky outcrops and the laterite crusts constitute the higher areas, whereas the valleys with clay-loam soils form the lower parts. Slopes of varying degrees, with intermediate soils, make the transition between the two types. The rainwater runs in streams down from the heights to the valleys where it sometimes forms temporary lakes. The intensity of run-off varies with the topography and the texture of the soil, but in general only 75% of precipitation penetrates the soil (Penning de Vries and Djiteye, 1982).

VEGETATION

Two principal types of vegetation are found in the region. The northern part (Sahel) is the area of steppe and shrubland, made up essentially of annual grasses and woody plants of the genera *Acacia*, *Balanites*, *Ziziphus*, etc. In the south, savannah replaces the steppe; the grass canopy improves with the appearance of tall perennial grasses. The woody vegetation becomes more and more diverse and dense as one goes south.

The steppe is an open formation. The rate of recovery of the herbaceous stratum is poor; the xerophytic character of the grasses is very pronounced: short growing season grasses with narrow leaves in circles or basal rosettes. Depending on the density of the different elements, the steppe takes the form of trees, shrubland or grassland. The spatial distribution of these different forms is a function of the nature of the soil and of the topography. The sand dunes, with poor water holding capacity, and the slopes rarely have woody plants but the herbaceous stratum, made up of annual grasses, is well developed.

The most common species are _Cenchrus biflorus_, _Schoenefeldia gracilis_, _Elionorus elegans_, _Borreria_ spp., etc. The importance of the herbaceous stratum is a function of the development of the trees and shrubs. The perennial grasses with longer growing seasons which appear include _Andropogon gayanus_ and _Diheteropogon hagerupii_. The transition zone between a dune area and an adjacent valley is generally formed by a compacted soil, impermeable and unsuitable for the establishment of vegetation. The last type is found locally between the dunes and valleys, separated by bare strips where only a few species with very short cycles (_Zornia_, _Dactyloctanium_) are still able to survive. The poor nature of the soil and the irregular rainfall impose an important restriction on production.

Penning de Vries and Djiteye (1982) estimate the total herbaceous biomass production of this region as 1000-2000 kg dry matter per hectare. Although the energy value of this biomass is satisfactorily maintained up to the middle of the dry season, it must be emphasized that the nitrogen content becomes insufficent from the end of the rainy season, with the flowering of the grasses. The contribution of edible material from the forage trees in this period (leaves, flowers and pods) partially compensates for the poor quality of the herbaceous biomass. Nevertheless, many areas remain unexploitable because of the lack of water.

Like the steppe, the savannah can take various forms: grassy savannah, shrubby, woody and forested. The dominant species in the southern part of the semi-arid zone are the annual grasses with long cycles: (_Pennisetum pedicellatum_, _Andropogon pseudapricus_, _Diheteropogon hagerupii_) and the perennial grass _Andropogon gayanus_. The biomass production of the grass canopy reaches an average of 3.5 tonnes per hectare. The extent of this production explains why this zone can burn each year during the dry season. The bush fires destroy the reserves of dry grass, reduce the seeds and modify the form of the trees. As a general rule, the areas in the north part are of better quality than those of the south, where in addition, the tsetse fly appears in the region of the 14th parallel.

The semi-arid zone of West Africa is crossed by two of the major rivers of the continent: the Niger and the Senegal. Along the route, these rivers overflow their banks and flood entire regions in which important aquatic grasslands develop. In the _Echinochloa stagnina_ flood plains of the central delta of the Niger, biomass production can reach 6-17 tonnes per hectare (Boudet, 1975).

The stocking capacity of the pastures of the semi-arid zone is strongly related to the rainfall (northern part) and the pattern of fires (southern part). Boudet (1975) calculated it at 50-60 kg/ha/year in the steppe and 80 kg on average in the savannas.

Furthermore, the state of the areas of the semi-arid zone is subject to major variations. If during the rainy period (3-5 months/year) the animals have at their disposal relatively abundant and good quality forage, the situation rapidly reverses in the dry season; in the north, pastures which are still of acceptable quality have to be abandoned through lack of water, just as in the south, a great part of the biomass is wiped out by the fires.

THE SYSTEMS OF PRODUCTION AND THEIR CONSTRAINTS

The semi-arid zone of Africa supports 60 to 70% of the national cattle populations of the countries concerned. Except for some establishments of an experimental nature (state ranches, pilot farms, etc.), livestock management is conducted according to traditional systems. Two main systems of production can be identified:

The pastoral system

This system is found in the northern part of the zone where the irregularity of the rains prevents the establishment of crops. The principal product is milk and the main function of the cattle is to more or less provide for human subsistence. The pastoral system is characterized by temporal and spatial mobility. The periodic migrations here take the form of nomadism between water holes in the dry season. The livestock involved are relatively few (less than 10% in Mali).

The herds are composed of cattle, small ruminants (notably goats) and camels, all contributing to milk production. In addition, the small ruminants provide meat to the herders and the camels are used for transport. Crops (mainly milo) provide important complementary foods; the herders obtain this in exchange for cattle or milk from the settled population. The level of commercialisation of the cattle is minimal, but occasionally two or three old bullocks or infertile cows can be sold in order to meet tax obligations or for the purchase of consumable goods. As well as their economic functions, the livestock dominate all aspects of social life.

Jahnke (1984) identified three fundamental principles of herd management in the pastoral system:

- adaptation to the natural environment: the availability of water and the quality of pastures determines the migration of the total or part of the herd over distances and times in different years. The overall size of the herds and the distribution of species have the objective of guaranteeing regular provision of milk by the best utilization of available vegetation;

- the prevention of risks: the animals are divided up into different groups in order to divide the risks of disease and to adapt to the requirements for distant pastures in case of prolonged drought. The tendency for the herders to increase numbers and keep old females is also part of this same strategy.

- adaptation to the institutional environment, characterized by a method of collective responsibility for pastures; increasing production by increasing numbers results from collective exploitation of pastures.

The productivity of livestock in the pastoral system is low (see Table 1), varying from one region to the other and between years. Camels and small ruminants are 1.5-2 times more productive than the cattle; but, on average, the annual productivity of the area varies from 26.2-31.4 kg/ha for milk and 2.6-3.1 kg/ha for meat (Jahnke, 1984).

Since the drought which started in the 1970s, there has been a major disturbance of the pastoral system. Faced with environmental degradation, an effort to adapt has resulted in certain cases in the settlement of some of the herders around permanent waterholes. At the same time, the distances involved in transhumance have been extended for the others. The proportion of cattle has been reduced in favour of small ruminants and camels, but overall, the equilibrium between present numbers and available forage resources remains precarious.

Table 1. The productivity of livestock in the pastoral production systems of tropical Africa.

Production	Cattle	Camels	Sheep & goats	Mixed herds
Milk				
kg/head	66	248	22	-
kg/livestock unit	95	248	220	161
Meat				
kg/head	9.6	7.4	3.5	-
kg/livestock unit	13.7	7.4	34.5	16.3

The agropastoral systems

These are the systems in which livestock production is practised in association with agriculture. This association may be close and complex, or livestock and crop production may, on the contrary, be parallel activities in their own right and can even belong to different management units. In this case, the association comes down to geographical proximity. Two principal sub-systems of management can be distinguished according to the nature of cultivation practised: rain-fed farming (mainly milo) and irrigated farming (rice).

In the rain-fed mixed farming system, a minimum of agricultural activity is carried out near to the dwellings, to provide crops for family consumption. The milk also contributes to the food supply but may also obtain some (variable) income for the farmer. Within the same production unit, the livestock can provide the means of production as animal power and manure. They generally benefit from the crop residues and, more rarely, from agro-industrial by-products whose distribution is limited to draught animals.

Competition exists, however, between the two activities in the use of land; it is accepted that it is the crops and not the livestock which provide the principal basis for subsistence. The latter do not usually benefit from the right to graze during the cropping period. The migration of zebu herds towards the north occurs regularly and is sometimes seen as inevitable. At the southern limit of the region where the trypano-tolerant animals are found, this transhumance is not generally practised.

The irrigated mixed farming system applies to the central delta of the Niger and of the Senegal. The income derived from livestock rearing is relatively limited. Transhumance takes place, as a rule, in the rainy season towards the Sahel, but a milking herd of females remains permanently near the dwellings to provide milk. It is in the dry period, after the wetland grazing and crop residues have been used up, that the greater part of the herd descends towards the semi-humid zone to the south in search of pasture.

The existence of belts of traditional dairies around the big towns of the semi-arid zone is well known. With the growth in demand for milk and the progressive transfer of animals from the hands of the traditional herders to the new breeders (businessmen, civil servants, etc.), a rapid evolution is taking place among these establishments. Production of milk for the market and the high demand for rentable land are profoundly modifying the feeding techniques. There is an increased tendency to keep exotic breeds with higher potential and massive utilization of agro-industrial by-products takes place.

The productivity of livestock in the agro-pastoral systems varies greatly from one region to the other. As a general rule, however, it remains low, but output is still higher than that of the pastoral system. In all situations, it appears that the genetic potential of the animals is not achieved in the traditional system. The production of milk and meat is clearly inferior to that obtained from the same animals placed in controlled management conditions with improved nutrition and health.

With the drought of the last two decades, the evolution of agro-pastoral systems has not favoured livestock production; the wetlands have been transformed into rice paddies and entire regions that were traditionally devoted to pasture have been put into crops. The weakening of the traditional policies for utilizing the resources and the absence of pastoral codes in the different countries has led to poorer and poorer management of pastures, notably around the towns and in the flood areas.

STRATEGIES FOR IMPROVING PRODUCTION

The improvement of feeding systems aimed at increasing dairy production in the zone should be based on:

1. In the pastoral system of the northern area:

   - a proper balance between the stocking rate and the carrying capacity of the pastures;

   - a rangeland water development policy that respects requirements for rational rangeland management;

   - improving impoverished rangelands by restoring vegetation where it has died out and by enriching cover with, for instance, legume crops;

   - controlling major epidemic diseases to overcome the herders' obsession with disaster and pave the way for new attitudes in favour of smaller, more productive herds.

2. In agropastoral systems and peri-urban dairying:

   - the preparation of pastoral codes that describe grazing rights in order to curtail agricultural encroachment and allow for better resource management;

   - closer integration of agriculture and livestock production through the introduction of fodder crops, such as cowpea, or the application of more fertilizer (including manure) to increase crop yields and, consequently, crop residues that can be fed to animals;

   - better use of locally available agro-industrial by-products (bran, polishings, molasses as urea-molasses blocks, etc.) through the preparation of supplements formulated to meet the needs of the animals and adapted to rangeland conditions;

- gradual intensification in peri-urban livetsock development projects to meet market needs. More importance should be given to experiments and the development of fodder crops as such, wherever conditions allow.

Any improvement aimed at increasing dairy production should consider the sociological aspect, which is closely related to livestock in this region. Some solutions have been proposed here to increase the milk production in the different production systems.

REFERENCES

Boudet, G. Inventaire et cartographie des pâturages en Afrique de 1975 l'Ouest. In: Inventaire et cartographie des pâturages tropicaux africains. Actes du Colloque de Bamako (Mali). 3-8 Mars, 1975.

Jahnke, H.E. Systèmes de Production animale et Développement de 1984 l'Elevage en Afrique tropicale. Kieler Wissenschafts-verlag. Vauk.

Penning de Vries, F.W.T and Djiteye, M.A. La productivité des 1982 pâturages sahéliens. Une étude des sols, des végétations et de l'exploitation de cette ressource naturelle. Wageningen.

FEEDING SYSTEMS AND PROBLEMS IN THE INDO-GANGES PLAIN: CASE STUDY

by

V.C. Badve

India has the largest bovine population in the world, with 191 million cattle and 69 million buffaloes, of which 80-85% of animals are nondescript. The numbers of milch animals are 50.7 million cattle and 28.3 millions buffaloes (Livestock Census, 1982). The National Commission on Agriculture, in its report in 1976, mentioned an annual milk yield of 157 kg and 504 kg from cows and buffaloes respectively. The low productivity of Indian animals is attributed to inadequate availability of feeds and fodders. The annual requirement of feeds and fodder are estimated to be 25.4 million tonnes concentrates, 353.0 million tonnes dry fodder and 308.1 million tonnes green fodder. However, only 16.5 million tonnes concentrates, 300.5 million tonnes straw and 261.0 million tonnes green fodder are available. The gap between availability and requirement of feedstuffs is wide, resulting in a large scale shortage. The occurrence of drought and flood has become a constant feature in most parts of the country. This creates serious problems with respect to livestock feeds.

In spite of all these problems, milk production in the country is showing an increasing trend. Currently, milk is the second most important agricultural commodity after rice. A decade ago milk production was hardly 30 million tonnes, while it reached 43.9 million tonnes in 1986-87. It is expected to exceed 50 million tonnes by 1990. The target set for 2000 AD is 65 million tonnes (Tables 1 and 2) (Chatterjee and Acharya, 1987). The increase in production is due to a massive cross-breeding programme, especially in cattle, and the use of improved quality feed and fodder. The population of crossbred cattle is estimated to be 12-13 millions, of which more than half would be breeding females with 1800-2400 kg milk production per lactation.

CURRENT PATTERN OF UTILIZATION OF FEED RESOURCES

Crop residues

Crop residues and other cellulosic materials are staple feeds for dairy animals in India. The most abundant residues are cereal straws, sugarcane tops, sugarcane bagasse, pulse straws, millet straws, etc. According to Singh and Rangnekar (1986), the availability of dry fodder/straw from grain and groundnut crops is estimated to be 302.39 million tonnes (Table 3). These feeds are unable to meet the maintenance requirement of animals because of the low digestibility, influenced by high fibre, lignin and silica content.

Table 1. Trends in milk production, *per capita* availability, processing and milch animal population since 1951 and projected to 2000 AD.

YEAR	HUMAN POPULATION (IN MILLIONS)	MILCH ANIMALS PRODUCTION (IN MILLIONS) COW	BUFFALOES	MILK PRODUCTION (MILLION TONNES)	PER CAPITA AVAILABILITY (GRAM /DAY)
1951	361	46.37	21.01	17.4	132
1961	439	51.01	24.24	20.4	127
1971-72	546	53.41	28.61	22.5	112
1981-82	685	54.37	28.65	34.5	136
1983-84	720	-	-	37.1	141
1985-86	751	55.45	33.07	42.3	154
1986-87	766	-	-	43.9	157
1989-90 (Projected)	812	-	-	51.0	172
2000 AD (Projected)	986	51.25	30.59	65.0	180

Source: Dairy India (1987).

Table 2. Trends in milk production by region in Sixth and Seventh Plans (production in million tonnes)

Region	1984-85 Target	Achieved	1986-87 Anticipated achievement	1987-88 Target	1989-90 Target (%)
1. Northern Region	18.17	17.77	19.53	20.56	22.96 (44.8)
2. Central and Eastern Region	7.69	8.43	9.42	9.79	11.27 (22.0)
3. Western Region	4.73	5.48	5.73	5.94	6.23 (12.16)
4. Southern Region	7.61	8.49	9.22	9.64	10.79 (21.04)
Total	38.20	40.17	43.90	45.93	51.25 (100)

Source: Dairy India (1987).

<u>Regions referred to in Table 2.</u>

Northern Region: Harayana, Himachal Pradesh, Jammu and Kashmir, Punjab, Rajastan and Uttar Pradesh.

Central and Eastern Region: Arunachal Pradesh, Meghalaya, Sikkim, Mizoram, Nagaland, Madhya Pradesh, Bihar, Orissa, West Bengal, Assam, Manipur, Tripura.

Western Region: Goa, Gujarath and Maharashtra.

Southern Region: Andhra Pradesh, Karnataka, Kerala, Pondicherry, Tamilnadu, Andaman and Nicobar.

Table 3. Estimated availability of dry fodders/straw from grain and groundnut crops in India in the year 1983-84.

Crops	Area million ha.	Production million tonnes	Ratio Grain/straw	Estimated dry fodder, million tonnes
Paddy	40.99	89.57	1:1.5	134.35
Wheat	24.39	45.14	1:1.5	67.71
Sorghum	16.26	11.93	1:3	35.79
Pearl millet	11.81	7.62	1:3	22.86
Finger millet	2.60	2.99	1:3	8.97
Small millet	3.61	1.71	1:3	5.13
Maize	5.88	7.92	1:2	15.84
Barley	1.37	1.78	1:1	1.78
Pulses	23.41	12.65	1:0.5	6.32
Groundnut	7.64	7.28	1:0.5	3.64
Total				302.39

Source:- Singh & Rangnekar (1986).

As feed supplies to the animals are closely tied to the local cropping pattern, variation in feeding regimes are observed from region to region. In the Northern part of the country, wheat straw (bhusa) is more intensively utilised, while feeding paddy straw is common in Eastern and Southern regions and part of the Western region, particularly in coastal areas. Sorghum stovers are fed in the Central and Western regions and in parts of the Southern region. Feeding millet and pulse straw is also observed in certain localities.

In sugarcane growing areas of the country (part of Uttar Pradesh, Maharashtra and Gujarat), sugarcane tops are extensively fed to dairy animals during the harvesting season from October to May. During the summer months, they constitute the bulk of green material available to animals in these regions. A survey was conducted with farmers of different size categories and in three different seasons in three villages from Western Maharashtra which showed seasonal variation in the feeding of cane tops (Table 4). Maximum use of cane tops (up to 52% of total dry matter) was observed with small farmers in the summer season (Thole et al., 1988).

Table 4. Seasonal changes in forage availability (%) in different farmer categories (dry basis)

Farm size	Large			Average			Small		
Season	Summer	Rainy	Winter	Summer	Rainy	Winter	Summer	Rainy	Winter
MAIZE	4	7	1	19	19	6	14	1	0
SORGHUM	7	21	9	15	41	23	7	36	6
CANE TOPS	21	9	17	31	9	33	52	11	34
LUCERNE	17	15	11	21	15	10	4	4	2
GRASS	7	13	11	2	3	2	2	16	11
DRY	13	15	8	8	8	8	19	30	44

Source: Thole et al., (1988).

Feeding concentrates

The use of agro-Industrial by-products, either as individual concentrates or as a part of balanced concentrate mixture, is a widely observed practice all over the country. In places, farmers mix by-products with conventional feed ingredients like brans and oilcakes, sprinkle some water on the mixture and feed animals at the time of milking. Salt or mineral mixture are often added to such feeds. It has been generally observed that concentrates are fed only to lactating animals.

With the increasing popularity of dairying, compound feeds are being adopted. Because of organised milk collection through dairy cooperatives, the supply of inputs to farmers has been made possible. In 1977, total compound feed production was 400,000 tonnes and this had increased to 1.56 million tonnes in 1987 (Figure 1). Today, there are 71 feed mills in the private sector and 44 under dairy cooperatives, with a total capacity of 2.7 million tonnes/annum. In Operation Flood III, Programme 10, additional cattle feed plants with 100 tonnes/day capacity will be installed. To keep the cost of compounded feed down, the technique of least cost formulation needs to be adopted widely.

Figure 1. Trends in cattle feed production in India.

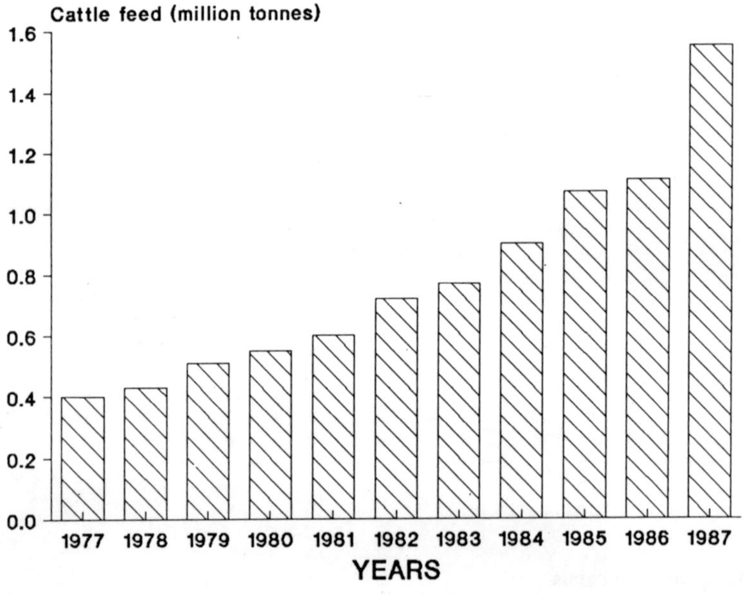

Feeding green forage

Although forage based feeding systems help lower feed costs, the scope for such systems is limited in India because of the need to give priority to food crops. The average cultivated area under fodder crops is estimated as 4.4%. In areas with better irrigation facilities, intensive fodder production is practised and in the

Northern Region, particularly Punjab and Haryana, 10% of the irrigated land is allocated to fodder cultivation. The major part of the ration of dairy animals in this region consists of lucerne, berseem, maize, oat, sorghum, etc. In other parts of the country, although the area under fodder crops has not increased, the technique of a mixed cropping system of forages with other cash crops like vegetables and sugarcane is widely adopted by farmers. Growing maize with sugarcane, brinjal, cabbage, etc., is a common practice in irrigated tracts of Maharashtra. Table 5 shows the crop rotation in different agro-climatic zones of India.

Table 5. Intensive fodder crop rotations for different agro- climatic zones of India.

ZONES	CROP ROTATION	GREEN FODDER YIELD T/ha/annum
NORTHERN	Hybrid napier intercropped with berseem	211.7
	Hybrid napier + lucerne	176.0
	Berseem + Japan rape - Jowar + cowpea	170.5
	Maize + Cowpea-Maize + Cowpea - Turnip-Oat	190.0
CENTRAL & WESTERN	Hybrid napier + cowpea-berseem + mustard	286.3
	Maize + cowpea-M.P.Chari-berseem + mustard	197.2
	M.P.Chari-turnip-oat	192.3
	M.P.Chari + cowpea-berseem + mustard-Jowar + cowpea	168.6
	Maize + cowpea- maize + cowpea-oat-maize + cowpea	168.5
EASTERN	Maize + cowpea-oat-bajra + cowpea	102.6
	Jowar + cowpea-berseem + mustard-maize + cowpea	96.0
	Maize + ricebean-berseem + mustard	111.5
	Hybrid napier alone	144.2
SOUTHERN	Sorghum + cowpea-maize + cowpea-maize + cowpea	110.7
	Maize + cowpea-maize + cowpea-maize + cowpea	106.0
	Guineagrass round the year	93.5

Source: Lal M. and Tripathi S.N. (1987)

Growing short duration forages in the gap period of the prevalent crop sequence is a standard practice in irrigated areas. For example in the wheat-sorghum-maize-bajra sequence, a gap period exists between April and June which is utilized for growing forage crop mixtures like maize + cowpea, sorghum + cowpea or bajra + cowpea with a yield of 35-40 tonnes/ha, without affecting main crop.

TECHNOLOGY OPTIONS AVAILABLE TO THE FARMERS

Supplementation

Supplementation of crop residues with fresh grasses and legumes or concentrate feeds significantly improves feed intake and the performance of animals. Feeding wheat straw with berseem or lucerne is common practice in the Northern region of the country. In dryland farming systems where forages are scarce, crop residues are supplemented with concentrate feeds. Supplementation of the basal diet with good quality forage or concentrates helps to overcome the problem of low palatability. The role of agro-forestry systems in augmenting the supply of green forage needs to be emphasized to farmers.

Urea treatment

Treating crop residues with 4 percent urea and 45-50 percent moisture improves the nutritive value by increasing digestibility, palatability and crude protein content. The process is simple and can easily be practised by the farmers. Feeding treated wheat straw supplemented with berseem (90:10 mixture on a dry matter basis) *ad lib*. was shown to support a milk production level of 6 kg/head/day without concentrates (Agarwal *et al*., 1988). However urea treatment is not yet used on a wide scale by farmers because of inadequate extension efforts to popularise the technology and the limited availability of liquid cash for farmers to purchase urea.

Steam treatment

According to Rangnekar *et al*. (1982, 1986), steaming under high pressure has been found to be effective for improving palatability as well as digestibility of sugarcane bagasse. It has been demonstrated that it is possible to utilise this process in sugar factories, since steam can be made available at a very low cost. Field trials with steam treated material have shown good results and acceptance by the farmers. In some areas, this material has been used as an alternative roughage source during feed scarcity to maintain animals while, in other areas, it has been used in complete feeds for lactating cattle.

Urea-molasses blocks

Urea-molasses blocks provide nitrogen to the micro-organisms in the rumen and thus improve the digestion of straw. They can also supply amino acids which can by-pass rumen fermentation and be absorbed in the lower gut of the animal. Cattle and buffaloes fed these supplements showed improved body condition, increased conception rates and increased milk yield. The National Dairy Development Board (NDDB) has launched a programme to popularise the feeding of urea-molasses blocks.

FUTURE DEVELOPMENT

In India, nutrition research should emphasize the development of feeding systems based on existing feed resources, under farm conditions.

A feed security system for animals needs to be developed to meet the requirements of livestock in famine and flood prone areas.

Evolving new varieties of cultivated fodders which have high yields, respond to inputs and are disease resistant is also a priority.

The identification of non-conventional feeds for livestock and developing processes for improving their nutritive values needs to be undertaken on large scale.

REFERENCES

Agarwal I.S., Verma M.L., Singh A.K. and Pandey Y.C. 1988 Studies on the effect of urea treated straw on milk production and feeding cost. Indian J. Anim. Nutrition. In press.

Chatterjee A.K. and Acharya R.M. 1987 Dairy Industry in India, a profile. In: Dairy India. pp. 3-19.

Lal Menshi and Tripathi S.N. 1987 Strategies and Technologies to maximise fodder production. Indian Farming, November 1987. p. 16.

Rangnekar D.V., Badve V.C., Kharat S.T., Sobale B.N. and Joshi A.L. 1982 Effect of high pressure steam treatment on chemical composition on and digestibility in vitro of roughages. Animal Feed Sci. Technol., 7:61-70.

Rangnekar D.V., Joshi A.L., Badve V.C. and Thile N.S. 1986 Studies on steam treatment of sugarcane bagasse for feeding dairy cattle: a review. Proceeding of International Workshop on rice straw and related feeds in ruminant rations, Eds. Ibrahim M.N.M. and Schiere J.B. Publ. Straw Utilization Project, No. 2, Kandy, Sri Lanka. pp. 192-200.

Singh K and Rangnekar D.V. 1986 Fibrous crop residues as Animal Feeds in India. In Rice straw and related feeds in Ruminants ration. Proceedings of an International Workshop held in Kandy, Sri Lanka 24-28 March, 1986. pp. 111-116.

Thole N.S., Joshi A.L. and Rangnekar D.V. 1988 Feed availability and Nutritional status of dairy animals in Western Maharashtra India A paper presented in "Workshop on Bioconversion of Crop Residues" held at Bangalore, 27-29 October, 1988.

FEEDING DAIRY CATTLE IN TROPICAL REGIONS OF CHINA

by

Cheng Nanging

INTRODUCTION

Guangdong Province is in the tropical zone of China with an annual average temperature of 19-26°C. The lowest monthly average temperature is 7-12°C and the warmest monthly temperature is 27-29°C. The amount of sunshine is 1600 to 2600 hours and accumulated temperature is 6000-9500°C. Average rainfall is 1000 to 2300 mm and non-frost days in 1986 where 346.

There are about 23,000 black and white cattle in the province, with about 12,000 (52%) on the state operated farms, about 5,000 on the collective farms (21%) and about 6,000 (26.2%) in the farmers' hands. There are about 2,000 water buffalo in milk, most of them belonging to farmers. Local water buffalo can produce about 800-1200 kg milk per year. The crossbred of local and Murrah or local and Nili-Ravi can produce about 1500-1800 kg milk and the highest can produce about 1800-2000 kg per year.

Besides selling fresh milk in the province, there are some milk processing plants. They produce yoghurt, vita milk, citrus fruit milk, condensed milk, etc.

In order to meet the people's milk demand, between 1980-1985, this province imported 3,318 black and white dairy cattle from New Zealand, Denmark, the USA, Canada and Australia. The milk productivity and management of these cattle and the original local black and white are as follows.

PRODUCTIVITY OF LOCAL AND IMPORTED DAIRY CATTLE

Adaptation to local conditions

Cattle from New Zealand are better adapted to the Guangdong Province conditions than cattle from other countries. In the first year of their importation, those cattle suffered serious heat stress. In the Chu-Cuen dairy cattle farm, among 675 cattle imported in 1985, 18.7% of them had to be slaughtered in the first year because of foot-rot, septicaemia, mastitis, pneumonia, dystocia, uterine prolapse, ruptured uterus and other diseases. In the second and third years, they had to cull only 5.8% and 1.9% of the total cows, respectively.

Milk production

From Table 1, it can be seen that the milk production of cattle from New Zealand is nearly equal to that found in their home country. The milk production of the cows from other countries, although it is increasing each year, is still lower than the production in their home countries (although higher than the New Zealand cows). The milk production has increased rapidly every year. Some of the cows produced 6000 kg of milk in a 305 day lactation and the highest one achieved 9327 kg of milk in 305 days.

Table 1. Milk production.

Country of origin	Lactation	No. of cattle	Average milk production kg	Average butter fat %
New Zealand	1	541	2948	4.3
	2	594	3580	4.3
	3	532	4003	4.3
	4	159	4257	4.3
	5	222	4187	4.3
USA	1	482	3013	3.2-3.7
	2	357	4280	-
	3	252	5405	4.4
Denmark	1	119	4434	3.9
	2	64	5123	4.0
Australia	1	6	4472	-
	2	3	4058	-
Guangdong black and white	1	984	4238	
	2	710	4797	
	3	656	5110	
	4	422	5146	

Reproductive performance

Table 2 shows that the reproductive performance of imported and local cattle are not very different.

Body weight of the cattle

The birth weight is smallest in New Zealand calves and their is not much difference between the others (Table 3). The body weights of heifers are highest in the American Holsteins.

Table 2. Reproductive performance.

Origin	Number of cattle	Conception rate at one heat %	Conception rate per year %
New Zealand	1132	57.2	91.7
Denmark	499	56.4	89.8
USA	1816	41.5	81.1
Guangdong black and white	1639	56.9	86.7

Table 3. Body weight (kg).

Origin	Birth weight	3 months of age	6 months of age	12 months of age	18 months of age
New Zealand	31.1	107.8	154.2	251.5	316.7
Denmark	35.6	-	138.9	267.9	340.9
USA	35.8	92.2	185.9	281.6	415.5
Guangdong black and white	34.5	-	-	-	345.7

GUANGDONG DAIRY CATTLE MANAGEMENT

Most of the dairy cattle are fed in-doors. In order to let the cattle adapt to the tropical climate and also to give high milk production, people have paid much attention to site selection, housing construction, dairy cattle management and disease prevention and cure.

Dairy cattle farm site selection

In the selection of farm sites, they have not only to consider those items such as prevention of sickness, transportation and communications, water supply, excrement and urine management and so on, but they specially select and construct a farm on the top of a hill, where the air flow is good and the manure, after fermentation in a manure pit, can be used to irrigate grassland around the farm automatically by gravitation through irrigation canals.

Housing design

In the past, most of the cattle houses had walls with doors and windows. Now, most of the cattle houses are open and they have only a big roof made of alloy, which is good for heat radiation. In windy areas, the roof may be made of cement.

Feed and feeding

Elephant grass is most important for cattle in Guangdong province because it can produce 8,000-15,000 kg of grass per Mu (1/15 of a hectare). Some of the farms plant a small area of stylo for calves. In the winter and spring season, they supply corn and elephant grass silage, and sometimes they also supply Chinese cabbage and sweet potatoes. Some of the farms supply grass hay the whole year round. In the concentrate, corn makes up about 40-50% and by-product feed ingredients such as wheat meal and soyabean meal make up about 30-50%. They also supply sufficient amounts of minerals, salt and some necessary trace elements. Most of the dairy cattle farms feed their cattle according to the feeding standards provided by the government. Sin-Tun dairy cattle farm uses a complete diet self-feeding system and gets very good results.

Dairy cattle management

The farmers do their best to avoid heat stress and foot rot in the cattle. Some of the farms lay bricks on the ground in the yard and some of them put sand down. Most of the cows' bedding is made of cement. Some of the farms have fixed bedding, with about 20 cm depth of sand in it, and some have a carpet on the bed to make the cattle comfortable. They put a water supply in the house, as well as outside, so that water is available all the time. Most of the farms have electric fans in the milking parlour as well as in the cows bedding area. Most of the farms let the cows have a shower once or twice a day and, in hot seasons, they use the shower two or three times a day.

Many dairy cattle farms are using milking machines to milk their cows. Chu-Cuen, Kwan-Ming and Sin-Tun dairy cattle farms are using herring bone milking parlours with fully automatic milking machines and the milking time per cow is only 8 minutes. They can machine-milk 12 to 20 cows at one time and have therefore raised their labour efficiency and produce very hygienic milk.

Management of bull calves

In many countries, young calves are sold for veal during the first few days of life but in our province, because the price of milk is very high and because most of the people cannot afford very expensive veal meat, most of the bull calves are slaughtered at birth.

DAIRY CATTLE PROJECT DEVELOPMENT AND INCREASED MILK PRODUCTION

Tentative ideas

In the past 10 years, thanks to the open door policy, the income of urban and rural people has increased remarkably and personal purchasing power has been raised in our country. The demand for more and more milk supply is a great problem for us to solve.

First of all, a better plan for dairy cattle distribution needs to be made and to increase the number of milking cattle. At present, most of the big dairy cattle farms are inside Guangzhou City or very close to Guangzhou. It is easier to transport milk to the consumers but there are many problems such as feed, especially forage and roughage supply, and pollution problems. So many people think that the distribution of dairy cattle should be changed.

a) Those farms in the city or very close to the city should raise a limited amount of good quality, high production cows, because they have a long dairy cattle-raising history, their technicians are good and so they could make good use of high quality cows.

b) Those farms in rural areas should raise cows that are better adapted to the conditions. In this area, it is easier to manage forage and roughage supply and the cost of labour is lower. With the improvement in transportation, it is not a problem to send milk to the city from 100 miles away in the rural areas. This is the area to develop more dairy cattle farms in the future.

c) There is a need to get more milk from water buffalo. In Nanhai and Jiashe County, there are about 2000 milking buffaloes. They produce about 2392 tons of milk per one year. Farmer who raise one buffalo for milk can earn 1200 - 1500 yen of per year. Our province has more than three million water buffaloes. If 5% to 10% are milked, it will not only provide more milk to the people but will also increase the economic income of the farmers.

From a genetic point of view, in order to raise more good quality, high production dairy cattle and have good quality bulls from those high production cows, we are trying to make use of embryo transfer techniques. In the past few years, we used Chinese-produced FSH to super-ovulate donor cows and obtained 5 - 6 viable embryos on average per cow. The transfer success rate of fresh or frozen embryos was about 40-50%. We are also planning to do some experiments using injection of recombinant DNA growth hormone products and try to make use to bio-technology methods to increase milk production.

MILK PRODUCTION SYSTEMS IN TROPICAL LATIN AMERICA[1]

by

J I Restrepo, E Murgueitio and T R Preston

INTRODUCTION

In many developing societies, cattle are more important as a source of manure - for fuel and/or fertiliser - and power, than of milk and meat. For the rural poor, they are more secure than the bank, as a means of safeguarding savings from inflation and devaluation.

In this paper, it is argued that it is more economical, in terms of national resource utilization, to satisfy the demand for milk and beef by combining both activities in the same animal. The justification for this approach is that: (i) the target levels of production - 2,000 litres of milk and 300 kg of beef per cow per year - are closely related to national demand ratios which vary from 4 to 5 litres milk per 1 kg of beef; (ii), as a result of (i), larger national cattle herds can be supported which increases employment opportunities and enables more efficient use to be made of presently under-utilised locally available feed resources, which are usually low in protein and high in cell wall material; (iii) advantage can be taken of important physiological traits, previously disregarded in intensive specialised systems - for example, the effect of suckling in stimulating milk yield, reducing stress in both cows and calves and permitting the calf to use supplementary feed of low protein content, more efficiently.

Of special importance to developing countries is that breeding programmes for dual purpose milk-beef systems permit a much greater degree of self-reliance (i.e. reduced dependence on expensive (imported) inputs), the technology is simpler and therefore more easily applied and with greater chance of acceptance than for specialised systems, especially milk production.

DUAL PURPOSE CATTLE PRODUCTION SYSTEMS

Dual purpose cattle production systems are those in which income is divided approximately equally between milk and beef. They are predominant in many parts of Latin America (see Table 1).

---

[1]Parts of this article were taken from "Dual Purpose Cattle Production Systems" by T.R. Preston and Lucia Vaccaro, published in "New Techniques in Cattle Production" (Editor, C.J.C. Phillips). Butterworths, London. Chapter 2: 20-32.

Table 1. Cattle production systems in the coffee-growing region of Colombia

	Altitude (m above sea level)			
	>2,000	1,250-2,000	<1,250	Total
Type of farms	%	%	%	%
Specialised beef	8	2	10	20
Specialised milk	9	9	2	20
Dual purpose	25	22	13	60

Source: Suárez and Jaramillo 1988

Their salient characteristics are that, almost invariably, the calves are raised on the cow by some form of restricted suckling. Usually milking is only once daily and the major feed resources are pasture or fibre-rich crop residues and by-products with minimum use of supplements.

The genetic resources vary enormously but the most popular animals for this system are crossbreds, derived from European *Bos taurus* types (Brown Swiss and Holstein predominantly) and *Bos indicus* (Zebu). Typical performance data from a number of countries are summarised in Table 2.

Table 2. Typical performance data for cattle managed according to the dual purpose system on demonstration or experimental farms in a number of tropical countries

	Milk/year		Weaning weight (kg)	Calving interval (days)
	Saleable (kg)	Calf (kg)		
Dominican Republic (1)	1,750	470	165	380
México (2)	1,400	450	150	401*
Costa Rica (3)	1,300	400	155	400*
Malaysia (4)	1,860	?	?	438*

*Exclusively with AI

(1) Fernandez et al 1978; (2) Alvarez et al 1980; (3) M E Ruiz, personal communication; (4) Cheah and Kumar (1984)

The dual purpose system arose through the need to increase the income from typically extensive beef production systems. Often the first stage is the milking of a proportion of the cows, those with appropriate genetic potential and temperament being chosen for this purpose. The next step is usually to introduce a sire from a recognised dairy breed, in order to increase dairy traits. Further innovations may follow, such as pasture improvement, supplementation of cows and calves, twice daily milking and occasionally machine milking.

More recently (Preston, 1977), dual purpose systems have been advocated as an appropriate way to integrate cattle into intensive mixed farms, especially in the wet tropics. The arguments used are that such systems enable better use to be made of available resources, that they are well understood by farmers (who developed them in the first place) and that they satisfy the demand ratio for milk and beef.

Aside from these economic considerations, there are distinct biological advantages intrinsic to dual purpose systems. These features are not well known and even less well understood. It is important to describe them, and what is known about them, so that those scientists that are in research centres in industrialised countries, who have the necessary laboratory resources and expertise, may feel stimulated to direct some of their attention to these areas with a view to establishing the underlying mechanisms.

RESTRICTED SUCKLING

Effects on the cow

Use of the calf to stimulate milk let-down is the traditional technique employed to coax beef animals to surrender a part of their milk output for human consumption. In crossbred cattle derived from Zebu (*Bos indicus*), typically used in dual purpose systems, there appears to be a negative linear relationship between the proportion of genes derived from the Bos taurus parent and the incidence of short lactations (Table 3).

Table 3. Effect of genetic makeup on incidence of short lactations in Holstein:Zebu crosses in México

Percentage of Holstein genes	Incidence of short lactations (<70 days) (%)
25	76
50	40
75	10
100	None

Source: Alvarez et al (1980).

In an unselected F1 herd (derived by crossing Zebu females with Holstein and Brown Swiss sires), milked by machine (Table 4), half the animals had lactations lasting less than 70 days when the calf was not present at milking. In their second lactation, those cows which had short lactations previously, milked normally when the calf was used to stimulate let-down. By contrast, the cows which milked normally in their first lactation (without calf stimulation), regressed to the mean in their second lactation, half of them becoming dry before 70 days.

Table 4. Milk production from F1 European (Holstein or Brown Swiss)/Zebu crosses milked with and without calf stimulation

	33 first-calvers	
First lactation	Without calf-stimulation	
	16 milked adequately	17 became dry <100d
Second lactation:	Without calf stimulation	With calf stimulation
Prematurely dry <100d	8	0
Lactation length* (days)	216	270
Total milk* (kg)	590	1680
Saleable milk* (kg)	590	1000

*For the cows which milked more than 100 days

Source: Alvarez et al., 1980.

As well as ensuring normal length lactations in crossbred cattle, restricted calf suckling brings other benefits. In a recognised dairy breed (e.g. Holstein), cows that suckled their calves after milking gave more milk during the period that suckling was practised and subsequently after the calf had been weaned (Table 5). There is less mastitis in cows that are milked and also suckle their own calves or calves from other cows (Table 6), compared with cows that are milked by hand or machine but do not suckle.

If cows which suckle their calves give more milk than those which do not suckle, it would be expected that either they must eat more food or mobilise more body tissue. However, in an experiment designed to test this hypothesis (Table 7), Holstein cows that suckled their calves after machine milking, gave more milk and lost less weight immediately after calving than cows which had their calves removed permanently 3-5 days after birth. The differences in body weight continued to be manifested at least through the first 3 months of lactation. Feed intake was maintained constant in both groups. The implication is that the stress on the dam caused by taking away its offspring led to adrenalin-stimulated demand for glucose and resulting increased mobilization of body reserves.

Table 5. Effect of two systems of restricted suckling on milk yield of Holstein cows and milk intake by their calves.

	Control (did not suckle)	Suckled	
		2xdaily for 70days	2xdaily for 28days then 1xdaily for 42days
Saleable milk (kg/d)			
5-28 days	12.5	9.7	9.5
29-70 days	11.5	9.5	13.5
71-112 days	10.0	11.8	12.9
Consumed by calf (kg/d)			
5-28 days	-	5.8	5.4
29-70 days	-	6.3	2.5
Total milk yield (kg/d)			
5-28 days	12.5	15.5	14.9
29-70 days	11.5	15.8	16.0
71-112 days	10.0	11.8	12.9

Source: J. Ugarte and T.R. Preston, unpublished data.

Table 6. Effect of suckling on incidence of sub-clinical mastitis (expressed as % of all quarters examined) in F1 (European x Zebu) and Holstein cows in the tropics.

Authors:	Breed	Calf suckling	
		No	Yes
Alvarez et al., 1980	F1(EXZ)	21	6
Ugarte and Preston, 1972	Holstein	6	2
Ugarte and Preston, 1975	Holstein	8	2

Table 7. Effect of suckling on milk production and bodyweight change in Holstein cows in Venezuela (The control cows had their calves removed permanently after the first 4 days; the experimental group suckled their own calves for 20 minute periods twice daily immediately after the cows had been machine-milked)

	Control (no suckling)	Restricted suckling	SEx
Milk production (kg/d)			
Saleable	7.9	9.0	±0.8
Consumed by calf	4.0	6.1	
Total	11.9	15.1	
Liveweight change (kg)			
Pre- to 7 days post-partum	-72	-46	±15
From 7 to 84 days post-partum	+15	+3	±5

Source: Velazco _et al_., 1982.

Table 8. Calves use milk more efficiently by suckling rather than by bucket feeding (calves were crossbred European x Zebu raised from birth to 84 days of age either by bucket feeding of whole milk or by restricted suckling for 20 minutes following milking.

	Bucket	Suckling	SEx
Condition Score*	1.61	1.35	±0.04
Milk intake (kg/d)	3.08	2.73	±0.12
Milk conversion (kg milk/kg LW gain)	9.7	4.9	±1.0

*Belly girth (cm)/liveweight(kg): low value = more tissue and less gut fill

Source: Fatullah Khan and T.R. Preston, unpublished data.

Effects on the calf

Efficiency of milk utilization is higher in calves that are suckled than when they take the same amount of milk from a bucket (Table 8). This is understandable in the light of Ørskov's work (Ørskov, 1983) which demonstrated that psychological stimuli, rather than physical factors, were the mechanisms which controlled the closing of the oesophageal groove which directs milk to the abomasum. Bucket feeding, by contrast, results in much milk spilling over into the rumen where the fermentative mode of digestion leads to losses in both the quality and quantity of nutrients available to the animal.

Other benefits are a reduced incidence of diarrhoea and elimination of navel sucking, as a result of which suckled calves can be housed in groups, permitting lower investment in housing, simpler feeding and management and less stress on the calves.

DISADVANTAGES OF RESTRICTED SUCKLING

Poorer fertility is generally ascribed to calf suckling, due to extension of the interval between calving and conception. It is generally believed that this is due to delay in initiation of ovarian activity. However, there is some evidence that the impaired fertility is due not to delay in ovarian activity but to poor manifestation of oestrus (silent heats) due to a reduced amplitude of the progesterone peaks which regulate ovarian cycles (Velazco *et al*., 1982).

Use of natural mating rather than artificial insemination is therefore advocated in dual purpose systems. This is substantiated by observations in a dairy enterprise in Mauritius where calves were raised by restricted suckling. When there was exclusive use of AI, calving intervals were long and variable; running bulls with the herd reduced both the average calving interval and variability (Naidoo *et al*., 1981).

RESTRICTED SUCKLING IN BOS TAURUS HERDS

Modifications to the management of cows and calves may be needed when calf suckling is introduced into herds in which the cows are mainly of Bos taurus origin and therefore do not need the physical presence of their calves to stimulate milk let-down. In such cases the calves are suckled when milking is completed, either in the shed where the cows are milked or in a pen designated for that purpose. It has been observed that, in this system, up to 20% of cows may withhold most of their milk during milking, retaining it for their calves. This problem can be overcome by cross-suckling, in a way which does not allow cows to suckle their own offspring (E. Murgueitio, unpublished data). For example, the cows in early lactation suckle the calves from cows in late lactation, and *vice versa*.

BREEDING PROGRAMMES FOR DUAL PURPOSE SYSTEMS

There is now broad agreement that in the humid tropics, the most appropriate animals for dual purpose systems are those derived by crossing native cattle (usually *Bos indicus*) with any of the recognised dairy breeds and that the optimum proportion of European genes will vary according to the harshness of the environment (McDowell, 1985). The results from the on-farm evaluations in Brazil, made by Madalena *et al*. (1982), show that there are few advantages and many disadvantages when the proportion of genes from a specialised European dairy breed exceeds 50%.

The most popular crossing sires are Holstein, Brown Swiss, Normandy and Simmental. Few breed comparisons have been made but the more reliable data indicate a significant advantage to the use of Holstein sires compared with Brown Swiss (Vaccaro, 1984). While there are many who advocate the merits of the native Criollo breeds in Latin America, their numbers are small and there are almost no data which permit valid comparisons to be made with other breeds and crosses (Vaccaro, 1987).

It is frequently argued that it is difficult to stabilise a cattle population in order to maintain approximately equal proportions of *Bos taurus* and *Bos indicus* genes. However, in practice this is not a major problem, once it is accepted that the appropriate way is by using F1 bulls. The recommended system is to "manufacture" such bulls by crossing native "adapted" females with imported semen from progeny tested sires of the selected European breed (Vaccaro L., personal communication). Hardiness and fertility are ensured by selecting the female parent for these characteristics. A sustainable level of milk production (1,000 to 1,500 kg per lactation) is guaranteed by choosing semen from a bull with a proven capability to maintain yields (in purebred dairy females) of about 5,000 kg per lactation (the potential yield of the F1 offspring is then at least 2,500 kg, ignoring both the dam's contribution and the effects of heterosis). Almost all dairy bulls presently standing at approved insemination centres in the industrialised countries have this capacity. F1 bulls can be run with the herd which facilitates natural mating. This is recommended in view of the difficulties of heat detection in cattle that raise their calves by restricted suckling systems.

CONCLUSIONS

The basic justification for the dual purpose concept is that the target levels of production - 2,000 litres of milk and 300 kg of beef per cow per year - are closely related to national demand ratios which vary from 4 to 5 litres milk per 1 kg of beef. The total cattle population required to support these yield levels is no higher than if the milk and beef were produced in separate herds, but with the additional benefits (for most developing countries) of supporting more employment opportunities and enabling greater and more efficient use to be made of presently under-utilised local feed resources.

Another important issue is that advantage can be taken of important physiological traits, which have been disregarded in intensive specialised systems - for example, the effect of suckling in stimulating milk yield, reducing stress in both cows and calves and permitting the calf to use more efficiently supplementary feed of low protein content.

Breeding programmes for dual purpose milk-beef systems are simple and low cost because they take advantage of F1 sires produced by combining imported "proven" (for milk!) semen with the adaptability and fertility of native females. This avoids the need to set up national progeny testing schemes which besides being expensive are also unreliable due to the difficulty of obtaining the necessary herd records.

REFERENCES

Alvarez F.J., Saucedo G., Arriaga A. and Preston T.R. 1980 Effect on milk production and calf performance of milking crossbred European/Zebu cattle in the absence or presence of the calf, and of rearing their calves artificially. Tropical Animal Production 5: 25-37.

Cheah P.F. and Kumar R.A. 1984 Preliminary observations on the performance of Sahiwal x *Bos taurus* dairy cattle. Malaysian Veterinary Journal 16: 1-7.

Fernandez A., MacLeod N.A. and Preston T.R. 1977 Production coefficients in a dual purpose herd managed for milk and weaned calf production. Tropical Animal Production 2: 44-48.

Madalena F.E., Valente J., Teodoro R.L. and Monteiro J.B.N. 1982 Milk yield and calving interval of Holstein-Friesian and crossbred holstein-Friesian:Gir cows in a high management level. Pesquisa Agropecuaria Brasileiro 18: 195-200.

McDowell R.E. 1985 Crossbreeding in tropical areas with emphasis on milk, health and fitness. Journal of Dairy Science 68: 2418-2435.

Naidoo G., Hulman B. and Preston T.R. 1981 Effect of artificial insemination or natural mating on calving interval in a dual purpose herd. Tropical Animal Production 6: 188.

Ørskov E.R. The oesophageal groove reflex and its practical
1983 implications in the nutrition of young ruminants. In: Maximum Livestock Production from Minimum Land (Editors: C.H. Davis, T.R. Preston, M. Haque and M. Saadullah) Bangladesh Agricultural University, Mymensingh pp.47-53.

Preston T.R. A strategy for cattle production in the tropics. World
1977 Animal Review 21: 11-17.

Suarez S. and Jaramillo C.J. Algunas caracteristicasde la explotacion
1988 ganadera en la zona cafetera de Colombia. Pasturas Tropicales 10: 24-27.

Ugarte J. and Preston T.R. Rearing dairy calves by restricted
1972 suckling once or twice daily on milk production and calf growth. Cuban Journal Agricultural Science 6: 173-182.

Ugarte J. and Preston T.R. Restricted suckling. VI Effects on milk
1975 production, reproductive performance and incidence of clinical mastitis throughout the lactation. Cuban Journal Agricultural Science 9: 15-26.

Vaccaro, Lucia El comportamiento de la raza Holstein Friesian
1984 comparada con la Pardo Suiza en cruzamiento con razas nativas en el tropico: una revision de la literatura. Produccion Animal Tropical 9: 93-101.

Vaccaro, Lucia Aspectos geneticos del programa de investigacion en
1987 bovinos de doble proposito en el ICA. Consultancy Report. ICA: Bogota, 26 pp.

Velazco J., Calderon J., Guevera S., Capriles M., Martinez N., Paredes
1982 L. and Lopez S. Efecto del amamantamiento restringido sobre el crecimiento de becerros, produccion de leche, actividad ovarica y nivel de progesterona en vacas Holstein y Pardo Suizo. In: Informe Anual Instituto de Produccion Animal. UCV, Maracay. pp. 66-165.

RESTRICTED SUCKLING IN DUAL PURPOSE SYSTEMS

by

J. Ugarte

INTRODUCTION

In the majority of the countries with a high level of agricultural development, the feeding of dairy calves is based on artificial rearing. The availability of milk substitutes based on by-products from the dairy industry has stimulated artificial rearing. This is not so in developing countries where there is a lack of fresh milk. In this case, artificial rearing must be based on fresh milk. Thus it is not logical to milk by hand or machine and later give part of the milk to the calf. It is better to make use of the calf's ability to extract milk from the cow.

Traditional rearing of dairy calves is characterized by the presence of the calf with the cow during milking to stimulate milk let-down and it stays with her after milking to consume the milk remaining in the udder. The time the cow is with the calf varies between 1 and 12 hours and is inversely related to the age of the calf. Age at weaning is rarely less than 6 months.

However, in dual purpose herds, natural (traditional) rearing does not allow efficient use of the cows' potential for the production of milk for the market. Hence, a variant of natural rearing was developed, called restricted suckling, characterized by the reduction in the time the calf remains with the cow each day, which is that strictly necessary for suckling, and in the age at weaning onto other feeds. This allows a greater economic effectiveness since:

1) it uses the maximum milk potential of the cows through the consumption by the calf of the residual milk.
2) it achieves high milk yields at milking and good calf growth.
3) it attains satisfactory reproductive performance and a low incidence of mastitis
4) it maintains a low calf mortality rate.

RESULTS OF RESTRICTED SUCKLING

Residual milk

15% of the milk in the udder at the start of milking remains at the end as residual milk, containing 3 times more fat content than normal milk (Lane *et al*., 1970).

The amount of residual milk has been correlated to total milk and with the interval between milkings. It varies throughout the lactation from 9.1 and 23.5% of total production in the first and tenth months of lactation respectively (Marx, 1971). Only small amounts of residual milk have been found in the udder of cows after the calves have suckled (1.2 - 3.4%), always less than in hand or machine milked cows (Kreilis et al., 1971).

Considering that the rate of milk secretion is more intense immediately after milking and that the amount of residual milk remains relatively constant, it is expected that, by delaying the suckling time, this recently secreted milk would also be consumed by the calf, with a possible reduction in the amount of milk to be obtained in the following milking. Milk consumption rose from 3.8 to 5.2 litres on increasing this interval from 20 min to 2 hours while production obtained at milking was reduced from 13.9 to 12.4 litres for a total production of 17.6 litres. By this method, the producer may obtain a certain amount of milk for the calf without affecting the total production of the cow.

The frequency of sucklings also affects the destination of milk produced. Thus, comparing suckling once or twice a day, Ugarte and Preston (1972) found that milk consumption did not differ between maternal breeds but was 50% greater for twice a day, while total production (milking and calf consumption) remained equal (Table 1). In the experimental cows, the decrease in milk production during the milking of Holstein cows suckling calves twice and once a day was 3.5 and 0.4 litres respectively, while in the F1 (Holstein x Zebu) it was of 3.6 and 1.2 litres. Suckling once or twice are equally effective methods of taking the maximum advantage of milk production in cows of this potential.

Table 1. Effect of suckling once or twice daily on milk production and calf growth (Ugarte and Preston, 1972).

	Litres of milk daily			
	At milking	Consumed by calf	Total	Daily gain kg.
Once daily x 60 m.	7.2	5.4	12.6	0.72
Twice daily x 30 m.	4.5	8.1	12.6	0.94
Control	8.0	-	8.0	0.54

A satisfactory variant was the combination of different numbers of sucklings (Ugarte and Preston, 1973) (Table 2). On reducing it from two to one from the fourth week of age to weaning (10 weeks), milk consumption was reduced by 54% (3 litres per day), as was the daily gain of the calf, although this was acceptable (478 g/day) and resulted in an average of 535 grams over the whole period 7 - 70 days. The production obtained at milking increased by 4.2 litres (31%), while the cumulative total to weaning was similar to the control (without calves) and total production was 28% higher.

Table 2. Effect of reducing suckling to once daily after the 4th week on milk yield and calf growth.

Milk yield (litres) 1-70 days

SUCKLING TREATMENT	1-28 days Milk -ing	To calf	TOTAL	29-70 days Milk -ing	To calf	TOTAL	1-70 days Milk -ing	To calf	TOTAL
2x daily to 70 days	9.8	5.8	15.6	9.0	6.8	15.8	9.3	6.4	15.7
2x daily to 28 days and 1x from 29-70 days	9.4	5.6	15.0	13.4	2.6	16.0	11.8	3.8	15.6
CONTROL	12.6	-	12.6	11.5	-	11.5	11.9	-	11.9

Milk yield (litres) 71-112 days by milking

2x daily to 70 days	11.8
2x daily to 28 days and 1x from 29-70 days	12.9
CONTROL	10.0

Daily gain of calves (kg)

	1-28	29-70	1-70	70-154
2 x daily to 70 days	0.735	0.954	0.865	0.705
2 x daily to 28 days and 1x daily from 29-70 days	0.624	0.478	0.535	0.718

It should be noted that the intervals between milkings were 15 and 9 hours and that restricted suckling took place after the milking with the shortest interval (afternoon), when the cows produced less milk. To increase consumption by the calf, suckling should take place in the morning, since differences of 0.8 and 1.0 litres were found on suckling the calves in the morning, compared to the afternoon, with intervals between milkings of 15:9 and 16:8 hours (Table 3).

Table 3. Effect of suckling after morning or afternoon milking on milk production and consumption by the calf (litres).

	At milking			Calf	Total
	Morning	Afternoon	Total	Consumption	
Experiment 1					
Morning	7.71	2.40	10.11	4.40	14.54
Afternoon	5.94	6.75	12.69	3.59	16.21
Experiment 2					
Morning	12.87	4.28	17.15	2.98	20.18
Afternoon	11.06	7.15	18.21	1.99	20.23

Age at weaning

This has usually ranged between 4 and 8 weeks. However, in systems of natural rearing, weaning ages have generally been high since it is not common practice to use sufficient amounts of concentrates. The calf must receive milk in the early stages to avoid seriously reducing performance.

When calves suckling once a day were weaned at 35, 56 and 70 days, no significant differences were observed in weight at weaning and at 154 days (Ugarte, 1977) (Table 4). On the other hand, total milk consumption increased with age at weaning. Total production (milking + consumption) over the period 7 - 70 days was apparently not affected by age at weaning. As particularly careful management is needed when calves are weaned at 35 days, 56 day weaning is recommended.

Table 4. Effect of weaning age on milk production and calf growth.

Period	Weaning age, days		
	35	56	70
Milk consumed (litres)	48	135	241
Net milk (litres)	1,210	1,242	1,088
Total milk (litres)	1,258	1,377	1,329
Liveweight gain of calves (kg)			
7 - 70 days	0.23	0.18	0.31
71 - 154 days	0.80	0.85	0.67
7 - 154 days	0.56	0.57	0.52

Milk production throughout lactation

A study carried out by Ugarte and Preston (1975) with 60 cows and calves suckling twice a day and an equal number without calves showed that from weaning (70 days) until cows were dried off, milk production did not differ significantly between treatments (6.1 and 5.7 litres per day, respectively). Lactation length did not differ either and total daily production throughout the lactation of cows suckling calves (milking + consumption) was 8.28 litres, while in cows without calves it was 7.35 litres (Table 5).

Table 5. Saleable milk, calf consumption and total production throughout the lactation (l/day).

	No. of animals	First 10 weeks Milk-ing	First 10 weeks To calf	First 10 weeks Total	From 10 weeks to drying off	Total	Lactation length days
Restricted suckling	57	6.2	6.9	13.1	6.1	8.3	262
Control	58	10.7	-	10.7	5.7	7.4	258
SE ±		0.4	-	0.3	0.2	0.2	8

These results agree with those obtained at a commercial level (with 97,678 cows milked twice a day for one year (Ugarte, 1977)). Daily milk production during milking was higher without the calf but, since about 332 litres were consumed by the calf, the F1 and F2 cows in restricted suckling produced 1.3% and 9% more milk in total (Table 6).

Table 6. Daily milk production (during one year) of F1 and F2 (Holstein-Zebu) rearing calves by restricted suckling or artificially (litres).

BREED	COWS			
	Without calf		With calf (restricted suckling)	
	no.	litres/d	no.	litres/d
F1 (50% H x 50% Z)	26902	5.91	54165	4.74
F2 (75% H x 25% Z)	9194	6.49	7417	5.73
Total	39096	6.05	61582	4.85

Mastitis

A study of 61 herds in tropical areas revealed that the incidence of mastitis represents 12.6% of milking cows (Fustes et al., 1985). A lower incidence of clinical and subclinical mastitis was found during the suckling period. After weaning, no significant differences were found between rearing systems (Ugarte and Preston, 1975) (Table 7). This is due to several factors, such as the mechanical effect of suckling, the cleaning effect of the saliva and a more complete emptying of the udder.

Table 7. Mastitis incidence in suckling cows.

HERD 1 Quarters affected by clinical and subclinical mastitis during the first 10 weeks of lactation (number).

	Cows	Clinical	Subclinical
Restricted suckling	36	5	14
Control	36	18	52

Table 7 (continued).

HERD 2	Quarters affected by clinical mastitis throughout the lactation %				
		Weeks after calving			
	Cows	1-10	11-20	21-30	31-drying
Restricted suckling	56	0.5	11	11	4
Control	58	2.0	14	11	4

Reproduction

It is accepted that natural rearing produces long calving intervals and this has been more evident on increasing the age at weaning. However, when the animals are early weaned (70 days) these considerations are not valid.

This was confirmed in a population of F1 and F2 (Holstein x Zebu) cows, without calves or suckling calves until 90 days (Table 8). Intervals between calving and conception were slightly longer with restricted suckling compared to artificial rearing, but considerably less than with traditional rearing. In another study, Rodríguez (1987), on analyzing more than 120,000 cows, reported intervals of 109, 151 and 218 days to first insemination and 128, 166 and 240 days to conception for artificial rearing, restricted suckling and traditional rearing systems respectively.

Table 8. Interval between calving and conception of cows under different management systems.

BREED	Number of cows	Artificial rearing	Restricted suckling[a]	TraditionaL[b] rearing
F1 (50% H x 50% Z)	27628	126 (8524)	166 (16461)	237 (2643)
F2 (75% H x 25% Z)	5680	136 (3179)	163 (2334)	178 (667)
Others[1]	54141	129 (17257)	175 (29245)	254 (7639)

[1]Mainly Zebu x Brown Swiss in different proportions
( ) Number of cows on each system
[a]Weaning at 3 - 4 months of age
[b]Weaning at 6 - 8 months of age

Mortality

On analyzing deaths occurring in 195,000 births during a one year period, values found were of 9.9%, 6.5% and 7.2% mortality for calves reared artificially, by restricted suckling and by natural rearing (Table 9).

Table 9. Calf mortality in different rearing systems.

Rearing system	No. of calves	Mortality %
Artificial	3820	9.87
Restricted suckling	75937	6.54
Other systems	116153	7.21

Cost of rearing

The cost of a calf reared artificially was 115.59 Cuban pesos, while, with restricted suckling, it was of 82.89 Cuban pesos, giving a difference of 32.70 Cuban pesos (Ugarte, 1977). Rodríguez (1987) also found a favourable difference for calves reared by suckling of 55.24 Cuban pesos (142.15 and 86.94 Cuban pesos in artificial rearing and suckling respectively).

REFERENCES

Fustes, E., Avila, C. and Ortega, L. Efectos de la mastitis bovina
1985 sobre la producción lechera y la economia agropecuaria. Rev. Salud Animal 7: 91-98.

Kreilis, M.L., Silinish, A.A. and Maksimova, E.P. The completeness of
1971 emptying of the udder with various methods of milk removed. Anim. Breed. Abstr. 39: 666.

Lane, G.T., Dill, C.W., Armstrong, B.C. and Switzer, L.A. Influence
1970 of repeated oxytocin injections on composition of dairy cows milk. J. Dairy Sci. 53: 427-433.

Marx, G.D. Use of a quarter milking machine to study factors involved
1971 in mastitis. Paper 7549. Scientific J. Serv. Minn. Agric. Exp. Stat.

Rodriguez, E. Incidencia de los sistemas de crianza de terneros en la
1987 actividad productiva y reproductiva en el ganado. Reunión ACPA. MIN. AGRIC. La Habana.

Ugarte, J. Crianza de terneros lecheros en amamantamiento
1977 restringido. Tesis Cand. Dr. Cs. ISCAH, La Habana.

Ugarte, J. and Preston, T.R. Rearing dairy calves by restricted
1972 suckling. 1. Effect of suckling once or twice daily on milk production and calf growth. <u>Rev. cubana Cienc. agric.</u> (Eng. ed.) 6: 173-182.

Ugarte, J. and Preston, T.R. Rearing dairy calves by restricted
1973 suckling. 3. The effect of reducing suckling frequency to once daily, after the fourth week on milk yield and the the growth of the calf. <u>Rev. cubana Cienc. agric.</u> (Eng. ed.) 7: 147-154.

Ugarte, J. and Preston, T.R. Rearing dairy calves by restricted
1975 suckling. 6. Effects on milk production, reproductive performance and incidence of clinical mastitis throughout the lactation. <u>Rev. cubana Cienc. agric.</u> (Eng. ed.) 9: 15-26.

HEIFER REARING IN THE TROPICS

by

J. Ugarte

INTRODUCTION

It is generally recognized that dairy heifers have lower growth rates throughout their life in tropical areas than in temperate ones (Figure 1). The pattern of growth is the traditional curved shape with a high growth rate and body development in the early stages of life, followed by a continuous slow increase as the animal gets older (Vaccaro and Rivero, 1985) (Figure 2). Liveweight differences between heifers reared in tropical and temperate areas are greater in animals over 18 months of age. These results suggest that more attention must be paid to this aspect, because of the positive relationship between mature weight and milk production.

These problems are not related to the genetic potential of the breed but are due to the environmental conditions, particularly feeding level (Menéndez, 1984).

PUBERTY

Age at first calving is basically determined by age at puberty. Average age at puberty is one of the most important components of the herd net reproduction because of its relation to the number of calves obtained each year and to the feed intake up to calving.

It is generally accepted that live weight is the most important factor affecting puberty. Heifers of large breeds usually reach puberty at 270 kg and smaller ones at 240 kg. In normally fed heifers, live weight is less variable and age at puberty tends to be relatively uniform, according to the breed.

LIVE WEIGHT GAIN AND AGE AT MATING

There is general agreement that it is economically advantageous to mate the heifers at early ages, no older than 15 months. This means daily liveweight gains from birth of 650 to 800 g. Higher gains of over 900 g/day, from 3 to 12 months of age, are undesirable because they decrease the growth of secretive tissue and increase fat deposition in the developing mammary gland. More difficulties at calving in over-fed heifers have also been reported (James and Tomlinson, 1988).

Unfortunately these liveweight gains and mating ages are not feasible in tropical areas because they depend on feeding cereals. Hence, it is necessary to look for alternatives that may not achieve

Figure 1. L..veweight of female Holstein calves.

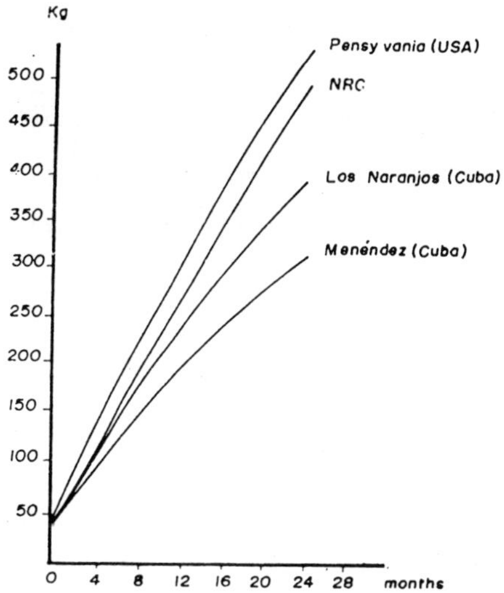

Figure 2. Liveweight of female Holsteins in the Venezuelan and Cuban tropics compared to Beltsville standards.

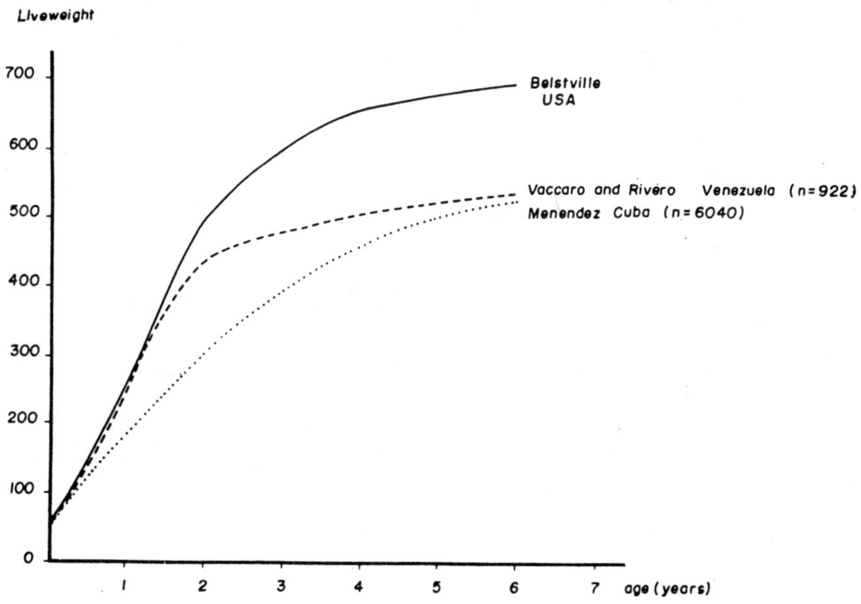

the performance obtained in developed countries but are adjusted to the prevailing conditions and naturally available resources of tropical countries.

It is well known that heifers in these areas are commonly underfed and receive poor management. Usually they graze on poor quality soils and receive small amounts of supplements or none at all. This could explain the low productivity and efficiency of cattle in such regions and the great number of unproductive animals in the herds (0.6 to 1.0 heifer/cow). Older ages at calving are mainly responsible for the latter results.

In the long term, insufficient feeding or seasonal scarcity of nutrients, affects reproductive performance in such a way that, even supplying high value diets afterwards, it is impossible to re-establish normal performance, even if the heifer weight is apparently high enough for normal reproductive activity (Perón 1984) (Table 1).

Table 1. Effect of nutritional level on reproductive behaviour (Perón, 1984).

	Nutritional level	
	Medium	Low[1] - Medium[2]
Puberty		
Age (days)	595	764
Weight (kg)	290	278
Average daily gain (g)	443	295
Progesterone level (ng/ml)		
Before puberty (35 d.)	1.06	0.77
During oestrus cycle	7.42	5.04
Feeding cost/heifer ($)	68.08	193.76

[1]406 days at a live weight gain of 173 g/day
[2]118 days at a live weight gain of 525 g/day

This latter author used low nutritional levels in 3/4 Holstein, 1/2 Holstein and 1/4 Holstein x Zebu heifers for 406 days and then a medium-high level for 118 days. He noticed that heifers reached puberty at higher ages and weights and the feeding cost was three times more than for those normally fed.

The effect of growth rate on age at first service and calving age was determined by Rosete and Zamora (1985) (Figure 3). They fixed 320 kg as service weight and daily gains varied from 350 to 600 g. For the highest gain, calving occurred at 27 months of age and, for the lowest, at 39 months.

The former age (27 months) is higher than previously reported for intensively managed herds in temperate areas but is more feasible to achieve in tropical conditions. 27 months is the age considered most adequate at calving for normally reared heifers (Ponce de León 1988).

The effect of age at first calving on total number of calvings is shown in Figure 4.

WEIGHT AT CALVING

It is not possible to analyze age and weight at first calving separately. Roy (1978) suggested different weights for Holstein heifers with different ages, according to the daily liveweight gain. He pointed out that liveweight before calving must be over 500 kg for 2 to 3 years old heifers.

Heifers must have good body condition at calving, with a high liveweight. This is reasonable, considering the relationship between weight at calving, milk production and weight changes in the first stage of lactation.

Low weights at calving are closely related to calving difficulties and subsequent reproductive disorders. This is the main reason for the high percentage of heifers which never reach the second lactation in tropical areas (48 to 63%).

UTILIZATION OF NATURAL RESOURCES

It is a fact that farmers in tropical areas must base their animal production on the utilization of natural resources, basically grasses and sugar cane, and on the agricultural and industrial by-products (Preston and Leng, 1987).

The possibilities of achieving an adequate weight (300-320 kg) and age (16-18 months) of heifers at mating on pasture were discussed by Zamora (1983). However, limited availability of irrigation and the high cost of fertilizers make it impossible to allow an adequate

Figure 3. Effect of growth rate on breeding age (Rosete and Zamora, 1985).

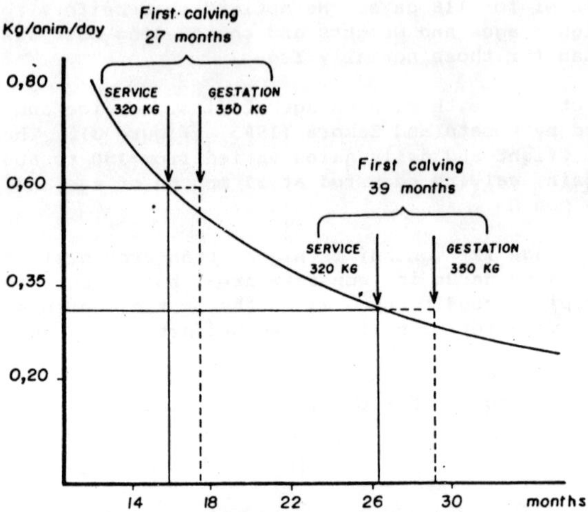

Figure 4. Effect of age at first calving on total number of calvings and abortions (Ponce de León, 1988).

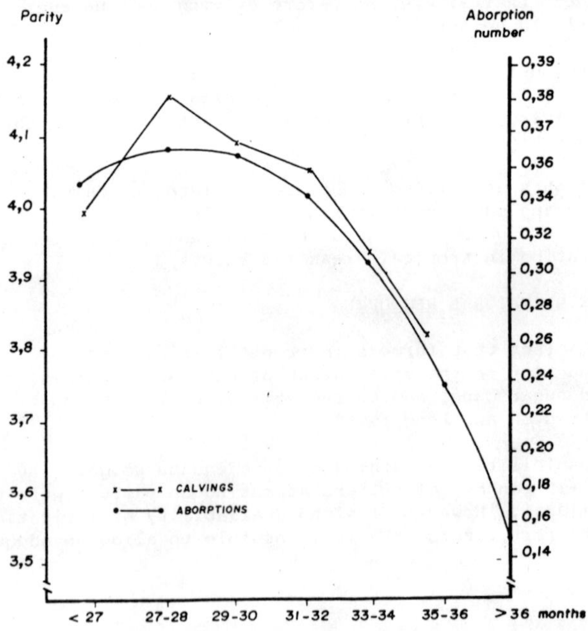

quantity of good quality pasture throughout the year. Hence, it is necessary to use other sources of nutrients in order to supplement the basic diet seasonally or throughout the year. There are large amounts of by-products and other materials that can be used for this purpose. By-products of the sugar cane industry are of major importance in most of countries, together with animal wastes such as poultry manure. Mixtures of these products, as supplements to pasture, have increased the heifers' daily gains to over 500 g (Perón 1984 and Rosete 1989).

On the other hand, sugar cane (whole plant) is successfully fed in the Caribbean area and South America but hitherto mainly with male calves, bulls and cows. In the future, heifer rearing could also depend on sugar cane during part of the year.

The use of legumes, fed as supplementary forage with pasture or sugar cane diets, has increased rapidly in the last 5 years. Relatively high weight gains (600 g/day or even more) have been obtained (Marrero, 1989).

It is not usually considered necessary to manage heifers separately from older cows in small dairy units. In medium and large dairy herds, poor body condition in heifers is frequently found, due to feeding competition. Increases of 10% of total milk yield (0.86 litres/ day), 6% less abortions, 7% less total animal losses and 1.4 months less in calving interval were reported when heifers were managed separately from older cows during their first lactation (Ribas et al., 1989).

REFERENCES

James, R.E. and Tomlinson, D.J. Debemos observar que tan rápido
1988 crecen las vaquillas. México-Holstein nov. 42-43.

Marrero, D. Sistemas de alimentación de gramineas y leguminosas para
1989 hembras bovinas de reemplazo en condiciones de secano. Tesis Cand. Dr. Cs. lCA, La Habana, Cuba.

Menéndez, A. Método simple para evaluar hembras lecheras. Rev. ACPA
1984 3: 15-19.

Perón, N. Efecto del nivel y tipo de alimentación en el
1984 comportamiento reproductivo de las novillas. Tesis Cand. Dr. Cs. Centro Investigación Mejoramiento Animal, La Habana, Cuba.

Ponce de León, R. Efecto de la edad al primer parto en lalongevidad y
1988 supervivencia del Holstein. Tech. dep.ICA. La Habana, Cuba.

Preston, T.R. and Leng, R.A. Matching livestock production systems
1985 with available resources in the tropics and sub-tropics. Penambul Books, Armidale, Australia.

Ribas, M. Pérez, B., Guzmán, G. and Mora, M. Informe Final de la
1989 etapa. Mimeo. Institute of Animal Science, La Habana.

Rosete, A. Contribución al estudio del valor nutritivo de la
1989 gallinaza y su aplicación a la alimentación de novillas lecheras. Tesis Cand. Dr. Cs. ICA, La Habana.

Rosete, A. and Zamora, A. Indicadores para la alimentación y
1985 manejo de novillas lecheras de reemplazo. Mimeo, ISCAH, La Habana, Cuba.

Roy, J.H. Rearing dairy herd replacements. J. Soc. Dairy Tec.
1978 31: 73-83.

Vaccaro, R. and Rivero, S. Growth of Holstein Friesian females in the
1985 Venezuelan tropics. Anim. Prod. 40: 279-285.

Zamora, A. Crianza de novillas lecheras de reemplazo en pastos
1983 tropicales. Tesis Cand. Dr. Cs. Institute of Animal Science, La Habana, Cuba

FEEDING COWS FOR MILK PRODUCTION IN THE ARUSHA/KILIMANJAROO COFFEE/BANANA BELT OF TANZANIA. FAO PROJECT: ASSISTANCE TO SMALLHOLDERS IN DAIRY DEVELOPMENT. CASE STUDY.

by

L.S. Morungu

INTRODUCTION

The feeding of dairy cows for milk production is a major problem in many tropical countries. The problem is even greater where limited areas of land are available for intensive pasture or fodder production, as is the case in the Arushu/Kilimanjaro coffee/banana belt of Tanzania. This paper seeks to describe the farming systems of the area in question and the initiation and implementation of a project to try to help dairy farmers in that area.

ARUSHU/KILIMANJARO REGIONS

Background

Arusha and Kilimanjaro regions, with a total area of 96,000 km^2, are located in the north-eastern part of Tanzania and are bordered by Kenya on the northern and north-eastern sides. The population of these two regions is estimated to be 2.6 million people, 80% of which live in the rural areas while 20% are found in the urban centres of Arusha and Moshi. There are three main agro-ecological zones:

i) The lowlands zone is characterised by unreliable rainfall and low population density. Drought is not uncommon and average annual temperatures are more than 30°C. The annual crops include maize, sorghum, cassava and, in the irrigated areas, one finds paddy and sugarcane. Extensive livestock keeping is practised in this zone the cattle grazed are predominantly, the Tanzania Shorthorn Zebu (TSZ).

(ii) The middle belt zone is characterised by a high populationdensity and a high rainfall, with temperatures ranging between 25 and 30°C. In this zone, one finds intensive farming of coffee, grown under bananas. While coffee is the main cash crop, bananas form the staple food. The lowest extreme of this zone is climatically similar to the lowlands and here one finds beef cattle production and the growing of annual crops, mainly maize and beans. In the coffee/banana belt, improved dairy cattle and also the (TSZ) traditional cattle are kept under zero grazing. This system is necessitated by the fact that farm holdings are very small (average 1 hectare) so that there is no spare land forgrazing. This belt is found on the slopes of Mounts Meru, Kilimanjaro and the Pare Mountain Ranges.

iii) The upper belt has very high rainfall, very high altitude and temperatures below 20°C. In this zone, there are national forests at the lower extremes and scant vegetation at higher altitudes. This zone is neither suitable for arable agriculture nor livestock keeping. This zone is mainly on the higher slopes of Mounts Meru and Kilimanjaro and to a lesser extent on the Pare Mountain ranges.

The project area covers a total area of 11,294 km², has a total population of 1,178,000 and a cattle population of 500,327 (of which 399,933 are indigenous and 87,197 are dairy cattle).

PRODUCTION SYSTEMS IN THE COFFEE/BANANA BELT

Categorisation of the farming systems has equipped the project to understand the farmers and their problems and in planning a strategic implementation programme. The main production systems are:

Category I. The farmer in this category has a small plot surrounding the homestead, mainly growing bananas. Sales of coffee are very low. In most cases, the farmer has no plot in the lowlands and, if any, it is difficult to cultivate. Capital availability is non-existent, due primarily to low coffee sales from his plot in the coffee/banana belt. The other characteristic is the absence of cattle, often coupled with the absence of small ruminants.

Category II. In this category, one sees a comparatively larger plot surrounding the homestead under bananas and coffee but the absence of a plot in the lowlands for annual crops and, if any, difficult to cultivate. Incomes in this category are mainly from the coffee sales while family labour only is available. The cattle found in this case are the local zebu (Tanzania Shorthorn Zebu - TSZ).

Category III. This is characterised by a piece of land around the homestead under bananas/coffee, as well as a piece of land in the lowlands, but difficult to cultivate due to lack of labour, distance and lack of or cost of transport to the lowlands. The source of labour is family labour, which in most cases consists of old people. Cattle are either the local cattle or up-graded dairy animals.

Category IV. This is characterised by a plot surrounding the homestead, also under bananas and coffee, as well as a plot of land in the lowlands. The source of incomes is mainly from the sales of coffee, milk and surplus bananas. The source of labour is the family, coupled with hired labour at certain times. The type of cattle are mainly dairy grade animals. There is no real accumulation of capital in this category.

Category V. This category is more or the less the same as Category IV above in terms of land ownership, source of income and labour, with the differences that there is sometimes accumulation of capital (purchase of land). Family labour is used part-time, including use of hired labour. The cattle are improved dairy cows.

Category VI. This category consists of farmers around or close to urban centres with very small plots. The source of income is from sales of specialised products like milk, chicken, pigs, etc. The source of labour is mainly hired labour, supervised by the family. Cattle, if any, consists of improved dairy cows.

INTEGRATED ASSISTANCE TO DAIRY DEVELOPMENT IN THE ARUSHA/ KILIMANJARO AREA

Since the inception of the FAO International Scheme for the Coordination of Dairy Development (ISCDD), missions were sent to a number of member countries who had expressed interest in developing or strengthening their dairy industry. Such a mission visited the Arusha/Kilimanjiro area in 1985 and recommended a dairy development programme based on an integrated approach, paying special attention to smallholders.

The problems of development are both of a technical nature and also constraints on inputs, facilities and services. The latter include feeding, breeding and AI; animal health and veterinary services; the dairy activities of the rural cooperatives - milk collection, processing and marketing; and also the training of both dairy farmers and technical personnel.

The programme was divided into nine sub-projects:

(1) Assistance to small holders in dairy development.
(2) Development of dairy activities in rural cooperatives.
(3) Fodder production on large scale farms.
(4) Expansion of the heifer breeding programme in Arushu/Kilimanjaro area.
(5) Strengthening of the veterinary services.
(6) Assistance to the vaccine production at ADRI in Dar-es-Salaam.
(7) Expansion of the AI services.
(8) Renovation of TDL operated collection/cooling centres and equipment of new centres at cooperative societies.
(9) Support to the rehabilitation of the TDL plant in Arusha.

ASSISTANCE TO SMALLHOLDERS IN DAIRY DEVELOPMENT - FAO PROJECT URT/86/013

This is the first of nine sub-projects and is currently being implemented and largely financed by FAO/UNDP; other donors include France, HPT, WFP, EEC, etc. It is a "transition project", due to the nature of its objectives (general and specific) and also because it lays the ground for the implementation of the other sub-projects. The general objectives of this project include strengthening of extension services, development of dairy activities in the rural cooperatives and thirdly the coordination of dairy development.

The specific objectives, which seek to deal with certain technical roles, include:

(1) to strengthen the extensive services of the regional and district livestock authorities in the project area, including farmers' training,
(2) to investigate and popularise the treatment of roughages with urea to improve its feeding value for smallholder farmers,
(3) to increase the quality and quantity of fodder grown by smallholder farmers,
(4) to increase the quality and range of inputs and services required by the smallholder dairy farmers, which are provided through the rural cooperative societies,
(5) to improve the nutrition of the animals by increasing the quality of molasses/urea mixtures fed,
(6) to procure suitable breeding bulls for use by smallholder farmers in areas where AI services are not readily available,
(7) to investigate the economics of baling roughages (especially maize stover) to reduce transport costs, and
(8) to assist in the coordination of various components of the integrated plan through support to the Dairy Development Coordination Committee.

PROJECT IMPLEMENTATION

Strengthening of the extension services

The extension services in Tanzania are centralized under the Ministry of Agriculture and Livestock Development. The scale of the project area, with 80,000 dairy farmers having a mixture of improved cows and milking zebu cows, makes it almost impossible to have an effective extension programme, considering the limited number of extension workers, especially when one thinks of individual visits as the sole extension method.

It was therefore advised that group activities, including meetings, seminars, demonstrations, study tours, farm visits and field days, be carried out with farmers. An intensive training programme was initiated for training local extension workers in two specific fields: specific technical messages and communication skills.

Involvement of women

The project has recognised the importance of women in dairy development. They are the ones that are attending the animals while the men are elsewhere. However, attendance by women at seminars has been minimal and this prompted the introduction of specific programmes (seminars) for womens' groups. Even though there has been some positive steps in this direction, more effort to involve women in extension seminars is being encouraged. To date, about 4% of those attending field days and seminars are women; contact is also made with women's groups in rural areas and also with other groups working with programmes which are in contact with the women in the villages.

Development of dairy activities in the cooperative societies

The rural cooperative societies are multi-purpose, with the main activity being marketing of cash crops, mainly coffee, and provisions of agricultural inputs to farmers. Since the farmers are also the cattle owners, the cooperatives have been encouraged to stock the inputs required by the dairy farmers, which include the concentrate feeds (wheat feed, maize bran, cotton seed cake and other cakes, etc.), molasses/urea mixture (MUM), dairy equipment, veterinary first aid kits and drugs.

The cooperatives have also been advised to form livestock committees to oversee this work, but this has not been effected to any substantial extent. Efforts have also been extended to advise the cooperatives to train their own personnel and, up to 1988, fifty-seven out of seventy MUM centres of the cooperative societies had sent their employees for training at the Livestock Training Institute, Tengeru. The cooperatives paid 50% of the course fees. Efforts are also under way to involve the cooperatives in AI field services for their farmers.

Dairy development coordination

The need for coordination arose from the fact that the project involves a lot of donors, including France, Britain, EEC, WFP, HPT, FAO/UNDP, etc. To coordinate all these donors and the different activities, the project has helped to form a Dairy Development Coordinating Committee which is mainly charged with coordination and monitoring of the implementation of the integrated dairy development plan.

The work of the committee is carried out by a series of technical sub-committees which in turn monitor the activities within their fields of competence. The sub-committees which have been formed under the coordinating committee include:

(1) Extension, Research and Training Sub-Committee
(2) Heifer Production and Distribution Sub-Committee
(3) Fodder Production and Distribution Sub-Committee
(4) Veterinary Services and Breeding Sub-Committee
(5) Women's Activities Sub-Committee
(6) Milk Collection, Processing and Marketing Sub-Committee

Feeding

Feeding is the main problem in the coffee/banana belt and this has been mainly due to scarcity of land. What the smallholder has tried to do to maintain his dairy animals has included the feeding of banana pseudostems, banana leaves, banana peelings, weeds and roadside grasses.

Except for the green feed which is grown on the edges of the coffee/banana plots, one will note that the rest of the feed resources are limited in terms of nutritive values which, in turn, affects DM intakes.

Efforts by the project have been directed towards increasing the quantity and quality of roughages produced by the smallholder farmers and also the introduction of legumes into the pastures and forage trees like *Leucaena* spp. Eight legume multiplication plots have been established and vegetative materials have been distributed to farmers.

Efforts to improve the feeding value of maize stover have been directed towards treatment with urea solution. Trials over the last three years have indicated that the method works and the extensionists have advised farmers to apply it by the pit method. Trials are underway to try large-scale treatment of maize stover at cooperative level. This will be undertaken, together with a study of ways of reducing transport costs of maize stover by baling, since maize stover and other roughages are transported from the lowland zone to the coffee/banana belt in loose form.

J.M. Centres has studied the impact of roughage treatment using the pit and basket methods (Table 1). Increasing the quality of maize stover at farm level has not been without problems (see below), despite the fact that practical guidelines were established for roughage treatment in terms of quantity of urea, maize stover, etc. Some positive aspects of roughage treatment have included increased intakes and particularly milk yields, and reduced wastage of the maize stover; 110kg of treated maize stover was enough to feed one cow for two weeks (with other feeds).

Table 1. Comparison of methods of treating maize stover with urea (J.M. Centres, 1989, unpublished data).

	PIT METHOD		BASKET METHOD	
	BEFORE	AFTER	BEFORE	AFTER
DM%	93.2	56.2	91.0	59.2
% DIGESTIBILITY	40.1	54.0	37.5	51.9
CP (%AM)	4.1	7.0	2.9	7.4

Feed Evaluation

This has been concerned with collection and evaluation of the feed resources currently used by the farmers. The feeds collected include banana leaves and pseudostems, elephant grass, guatemala grass, roadside hay, weeds from coffee plantations, bean straw, maize stover and even bean trash.

Description of the nutritive value of feeds used in the project area will provide information for adequate feed formulation. Determination of changes in nutritive value as a result urea treatment will provide information on the economics of treatment. Determination of differences in nutritive values of different varieties of maize stover and bean straw will help to advise farmers (and even plant breeders). Determination of optimum length of time for treatment of stover/straw with urea and optimum quantity of urea is being carried out under farm conditions. Lastly, measuring the changes in production of milk following introduction of new technologies would help to evaluate their impact.

The feed evaluation aspect has been done and will continue to be done jointly by the project, the Sokoine University of Agriculture and INRA-France.

PROBLEMS AND ACHIEVEMENTS

J.M. Centres (1988) has analyzed the data on extension. The number of seminars has increased 460% between 1986 and 1988, while the increase between 1987 and 1988 was 170%; 40,354 farmers have been contacted since 1986 through seminars/demonstrations. In 1988, only 18,773 farmers were contacted through 638 seminars/demonstrations. Assuming farmers attend more than one seminar or demonstration, the reports say that only around 15,000 farmers will have been contacted in three years.

This figure is low considering the total number of around 80,000 farmers in the project area. The frequency of the seminars/ demonstrations needs to be increased in the villages and adjusted in relation to the dairy cattle population. Efforts to get more farmers attending these sessions need to be promoted. Emphasis should be put on women involvement. Field staff involvement needs to be reviewed, together with more follow-up of the farmers after demonstrations. Timing of seminars/demonstrations should coincide with topics.

Extension Materials

To aid in the extension programme, a number of extension materials have been prepared, distributed and used both by farmers and extension workers.

<u>Handouts</u>. These have been prepared on 16 different topics including: <u>Desmodium</u>, <u>Siratro</u>, <u>Leucaena</u>, establishment of <u>Desmodium</u> cuttings, 'Grass equals Milk', feeding of dairy cattle, feeding of pregnant cows, molasses/urea mixture feeding, roughage treatment, dairy cattle breeding, milk production from zebu cattle, calf rearing, calf housing, housing of dairy cattle, milking hygiene, etc.

<u>Booklets</u>. These have been prepared and distributed. These include calf-rearing (6040 copies), dairy cattle breeding (4670), milking hygiene (1430), diseases of dairy cattle (3920), roughage treatment (1020), feeding of dairy cattle and women in dairy development.

<u>Slides and films</u> have also been prepared on the topics mentioned above.

Between July 1988 and March 1989, a total of 59 seminars and 115 demonstrations were carried out, including 31 with audio-visual support. Around 3800 farmers attended these seminars and demonstrations.

<u>Pasture establishment</u>

The main problem here has been seed availability. The project has been trying to combat this problem by using the limited seed available to establish pasture seed multiplication plots in different locations. It is hoped that, when this is achieved, there will be seed available to farmers. Another problem related to this is the problem of seed-setting with <u>Desmodium</u>. It is not known as yet why this legume is not producing seed. The project has therefore been distributing cuttings to farmers and the results are encouraging.

<u>Roughage treatment</u>

The project has conducted roughage treatment (mainly maize stover) campaigns at farmers' level (small scale), advocating the pit method mainly and the basket method since 1986. Technically, the pit method is a good one in terms of practicability, reduced wastage and increased yields. Problems encountered in 1986/87 have included moulding during treatment, storage of the treated product, lack or limited amount of maize stover, insufficient labour during treatment and transport of the maize stover in sufficient quantities. These problems have been compounded in 1987 by the interference of the Rinderpest Campaign, abnormal rain distribution and drought, and a shortage of the roughage treatment booklets.

J.M. Centres has also looked at the percentage of farmers who have repeated the roughage treatments after the one they did during the demonstration and also through attendance at the opening of the pits. The numbers of farmers using the treatments were 28 and 76 in 1986 and 1987 respectively, and the numbers repeating the treatment in

1987 and 1988 were 12 and 17.

Large scale treatment

This started this year, 1989, but on a very limited scale due to the rains. Baling was done on only 4 farms in the Rombo District due to distances between farms and the bad roads. It is expected that large scale baling and treatment will begin August/September, 1989.

Molasses/urea mixture feeding (MUM)

The idea of using molasses from the Tanganyika Planting Company (TPC), which is a sugar factory in the project area, goes back to 1976. The plans for the scheme, as it is currently being developed, were drawn up in 1980 but implemented only in 1982. At that time it was estimated that initial demand would be in the region of 3,000 tonnes per year with a long term potential of 10,000 tonnes. At that time molasses was plentiful and the project was assured of its short term requirements.

The mixing plant was completed in 1984 and the first sales of MUM were in 1985. Initially there were village tanks served by a tanker lorry. In 1986, plans were drawn up to increase the number of villages tanks from 20 to 70 and commissioning of the new tanks started at the end of 1986, and by 1986/7, 40 were in operation. New tanker lorries were also made available.

The initial price was 700 Tanzanian shillings (Tz) per ton. This was calculated on the basis of Tz 450/- for raw molasses and a further Tz 250/- for the urea and mixing charges. However when the tendering system was introduced at the start of the 1986/87 season, the price of MUM to the project was kept at Tz 750/- to help establish the use of this feed by dairy farmers. In mid-1987, the project was informed that it would have to pay Tz 3,500/- per ton and that the quantity allocated to the project was 3000 tonnes for the 1987/88 season. While the price to farmers had been 2.00 Tz per litre in the past it had to be 8.00 Tz for the same quantity. This had a very serious effect on the sales and use of MUM by the farmers in the project.

Experience to date shows that farmers cannot compete in terms of price with other users of MUM, even if demand was increased. Sales of MUM have therefore decreased due to prices.

The tendering system has therefore priced the MUM out of reach of many of the small scale farmers. Scarcity of other feeds makes MUM an extremely important component of the programme to raise milk yield. The project is still fighting for a price policy to be based on either the price of milk, the price of other livestock feeds (e.g. cottonseed cake, maize bran, etc.), TPC's costs of production or national inflation.

MILK MARKETING IN THE PROJECT AREA

Commercial milk marketing

In the project area is a milk processing plant situated in Arushu town, the Tanzania Dairies Ltd (TDL), which is one of seven in Tanzania. This plant collects milk from both large scale farms and smallholder farmers. It operates over 9 collection and cooling centres in the project area; the distance between TDL and the cooling centres varies from 30 to 100 kilometres.

The plant is faced with a problem that not all the cooling centres are in working order and therefore the plant does not work to capacity, as it cannot collect all the milk available and hence depends a lot on reconstituting and marketing reconstituted milk. The constraints that the plant is experiencing renders it ineffective in organising collection, processing and marketing of milk. The result has been vending of milk by middlemen and the establishment of dairy cooperatives in rural areas.

There are in the project are 3 dairy cooperatives in Kilimanjaro area where 2 are transporting and selling milk to Moshi town, while one is processing milk into cheese and markets it in Arusha and Moshi townships. These dairy cooperatives and other smallholders have found it a good business to sell the milk on their own, rather than to sell to TDL, because the price paid by the latter is lower than that which could be obtained by selling direct to consumers in town or their neighbours.

Commercialisation of milk has been growing over the years due to an increasing consumption pattern. This has resulted in smallholders investing in concentrate feeds, minerals, vitamins and drugs. This is due to high prices paid for milk and dairy animals, and the fact that farmers have realised for a long time that milk is nutritious and also an economic use of scarce land.

CONCLUSION

The feeding of dairy cattle for milk production in the Arushu/ Kilimanjaro area is the biggest problem, due to scarcity of land which necessitates zero-grazing and the use of diverse feeds from various sources. This requires, among other things, an efficient system to increase the quantity and quality of agricultural by-products and, most of all, an efficient extension programme. Involving the rural cooperatives in dairy activities will help farmers in obtaining the necessary dairy inputs locally. There is considerable potential for increasing milk production from the dairy and local cattle population, as the demand for milk is still far from being satisfied.

MILK PRODUCTION FROM TROPICAL FODDER AND SUGAR CANE RESIDUES CASE STUDY: ON FARM RESEARCH IN MAURITIUS

by

A.A. Boodoo

INTRODUCTION

The cattle population in Mauritius amounts to about 35,000 head, made up of various breeds. Of this, approximately 7,000 are females of producing age. The cattle industry in general can be described as being a low input system. The local cattle are called Creole; they are *Bos taurus* type and of medium size, 300-450 kg adult weight, polled and humpless. They are predominantly white or white-brown with dun, black or brown characteristic spots (Bennie, 1956). Cross-breeding programmes, using AI and bulls, have resulted in various levels of Friesian-Creole crosses. Other exotic breeds and their crosses exist in smaller numbers.

The small farms in the villages consist of 1 - 4 cows per household. These cows have a milk yield of 1200 - 1500 litres per lactation, short lactations (225 - 250 days), long calving intervals (15 - 18 months) and they first calve at 3 - 3½ years of age. The cows are hand-milked twice a day, generally before sunrise and at sunset. All cattle are kept indoors and fodder is brought to them. The stables vary from very simple ones built of poles with a thatched roof to improved ones with concrete walls and a roof of iron sheets.

Feed resources

The traditional practice of cowkeepers (small cattle owners in the villages) is to feed their cows mainly on sugar cane tops, which are abundant during the sugar cane harvest season (June to November), together with some selected grasses and crop residues. During the rest of the year, they feed a mixture of various grasses, creepers, shrubs, twigs and crop residues; these forages are available in varying amounts all the year round. Most of them are highly fibrous and contain 4 - 12% crude protein in dry matter. All forages are collected free from the neighbourhood and none are cultivated, at the cowkeeper level, for use as cattle feed.

Socio-economic importance of cowkeepers

The cowkeepers form an important socio-economic group as they supply about 95% of the fresh milk produced in the country. This is equivalent to about 12% of the total consumption of milk which amounts to about 90 million litres per year (fresh milk and imported milk powder). Cattle rearing in the village smallholdings is a family business and generally a part-time activity. This makes the business a flexible one in the sense that, depending on circumstances, the smallholder can add or sell one or two head of cattle quite easily. This is perhaps one important factor contributing to the fact that, despite the recent wave of industrialisation with its accompanying migration of labour from agriculture to the factories, the cowkeeper community has continued to be in business, although their number has decreased compared to a couple of decades ago. The cowkeeping tradition is still present in the rural areas and people still like to invest in this family business. There is a continuing demand for fresh milk both in the urban and rural areas.

BACKGROUND

In 1971, the Milk and Meat Project (FAO) diagnosed that lack of supplementation limited milk production. However, it was not specified whether it was energy or protein in the supplement that was important, and the basal diet of cane tops and grasses was not evaluated. This FAO study also proved technically that milk yield could be increased considerably by better feeding and management.

Recent research findings (Ma Poon <u>et al</u>., 1977, Gaya <u>et al</u>., 1982), obtained by the Ministry of Agriculture on Government farms, pointed out that one important limiting factor regarding milk production was the role of appropriate supplements which can stimulate consumption and utilization of roughages rather than depress roughage intake.

On the basis of these findings, the need was felt to further investigate the milk production potential of cows in the country. As the cows in the village smallholdings make a major contribution to the national production of fresh milk, it was decided, within the context of a project funded by the UNDP, to carry out the on-farm research described here in these smallholdings. This was decided because it was possible to have access to a large number of pregnant cows (experimental units) in a relatively short time and at almost no cost in the villages, whereas it would have been very difficult to obtain similar facilities on state farms.

OBJECTIVES

The objectives of this study were to:

a) investigate the effect on milk yield of supplementing the village cows with two types of concentrate, the traditional dairy concentrate (cowfeed), consisting of about 45% locally produced ingredients, and imported cottonseed cake;
b) describe the forage feed base at the cowkeeper level;
c) evaluate the performance of the local Creole breed, Friesians and their crosses.

ORGANIZATION OF ON-FARM TRIAL

Multi-disciplinary approach

The Animal Nutritionist was the team leader. The Extension Officer organized the evening meetings with the cowkeepers. Extension Assistants residing in the villages made daily visits to the farms and helped in data collection. They travelled on bicycles. Notebooks were kept at the cowkeepers' homes for recording animal weights, milk production, fodder offered, etc. The Veterinary Officer looked after the health of the cows and assessed pregnancy.

Equipment used

This consisted of a van for transporting staff, concentrates and an electronic cattle scale; a bathroom scale to measure birth weight of calves; a spring balance to measure the quantity of forage; and a kitchen scale for concentrates and minerals.

Specialized facilities

A chemistry laboratory was available for chemical analysis of feedstuffs and cannulated animals for nylon bag study of forages.

PROCEDURE

Concentrates

The cowfeed had 17% crude protein and was made up of 30% cane molasses, 30% cottonseed cake (or groundnut cake), 5% wheat bran, 11.5% rice bran, 20% maize, 1% common salt and 2.5% calcium carbonate. Cottonseed cake had 44% crude protein and was fed together with a mineral supplement of 15 g common salt and 50 g calcium carbonate per day. It was chosen for comparison with cowfeed because the Ministry proposed to use it later as a straight protein supplement, thus sparing mixing and transport costs.

Forages

The cowkeepers fed their cows forages *ad libitum* according to normal practice. Regular visits, 3 times per week, were made to the cowkeepers' farms and observations were made of management practices and animal behaviour associated with the supply and consumption of fodder. Measurements were made of total feed intake on 30% of the total number (88) of cows participating in the project.

The cane tops were first separated into two fractions, sheath bundle and leaf blade, with a large knife before analysis. For nylon bag work, two mature Friesian x Creole steers fitted with permanent rumen cannulae were used. They were fed a mixture of 20 - 25 kg

Setaria sphacelata and Ischmaemum aristatum and 1 kg cottonseed cake plus minerals.

Choice of cows

Cowkeepers who were willing to participate in the project voluntarily registered their names at their local Extension Office. When their cows were seven months pregnant, they started to receive the concentrates. Each cow was allotted in turn to either the cowfeed or cottonseed cake treatment.

Table 1 summarises the pattern of feeding the supplements which were given in two feeds daily.

Table 1. Daily levels of supplementation during the last 3 months of pregnancy and during lactation (Boodoo et al., 1988a)

	Cowfeed	Cottonseed cake
7th month (kg/d)	2	1
8th and 9th month (kg/d)	3	1
Lactation (kg/l milk)	0.5	0.25

The amount of cowfeed fed was twice that of cottonseed cake, because it had less than half the amount of protein as compared to cottonseed cake. Supplementation started at the end of pregnancy to make sure all the cows participating in the trial had adequate nutrition for fetal growth and lactation.

Management of calf

Only one aspect of the traditional management of the cowkeeper was interfered with during this study; anyone who used to allow his calf to suckle its dam was required to feed it from a bucket from the 7th day after calving until weaning at three months. This was necessary to measure total milk yield accurately.

RESULTS

The trial was conducted in 2 different climatic areas, with 3000 and 1450 mm rain per year respectively. The mean milk production of the cows (all breeds together) on the two types of supplement is summarized in Table 2. Milk composition was determined from monthly morning milk samples.

Table 2. Mean milk production and milk composition data, 22 cows per concentrate in each area (Boodoo et al., 1988a).

	Wet uplands		Dry Northern Area		Significance	
	Cowfeed	CSC	Cowfeed	CSC	Area	Feed
Milk production (kg)						
301 day lactation	3023	2871	2538	2649	P<.05	NS
SE	146	104	139	129		
Milk composition						
Mean fat (%)	4.08	4.57	4.31	4.61	NS	NS
SE	0.222	0.164	0.183	0.192		
Mean protein (%)	3.41	3.58	3.47	3.40	NS	NS
SE	0.149	0.118	0.061	0.089		

There was no significant difference between cowfeed and cottonseed cake in terms of milk yield. However, supplementation increased milk production by about 1400 kg per lactation, compared to the traditional yield of about 1200 - 1500 kg. Supplementation also prolonged lactation from the national mean of 225-250 days to 301 days. The lactation curves (Boodoo et al., 1988a) were of the classical shape with peak lactation (14.7 and 11.3 litres/day, for the wetter and drier areas respectively) occurring in the second month. The upland cows produced significantly more milk but there were no significant differences in milk fat and in milk protein due to the different areas and concentrates.

The effect of breed on milk production

The cows were classified on the basis of their phenotypic appearance into (1) the local breed, Creole (23 cows), (2) Creole x Friesian (47 cows) and (3) Friesian (18 cows). Table 3 summarises their milk production data. There were no significant differences in total milk production between breeds. These yields compare with national yields of up to 1500 kg for cows receiving little or no supplement and imply a response to supplementation.

Effect of harvest season on milk production

The average milk production of cows calving during the sugar cane crop season (June to November) was 2950 ±92 kg and was significantly ($p<0.001$) higher than the average for cows calving outside the cropping season (December to May) which was 2705 ±76 kg. There was not a statistically significant interaction between area and time of calving on the total milk production per lactation.

Table 3. Milk production data (kg) by breed (Boodoo *et al*., 1989). (SE in brackets)

		Creole	Crossbred	Friesian
Uplands	301 days	2788 (232)	2958 (115)	2899 (176)
	Peak[1]	13.0 (0.2)	14.5 (0.4)	13.0 (0.8)
Northern Area	301 days	2889 (216)	2536 (124)	2459 (156)
	Peak[1]	12.1 (0.4)	12.4 (0.6)	10.3 (0.7)

[1] Mean daily production in the 2nd month of lactation

OBSERVATIONS ON THE INTAKE OF FORAGES

The cane tops that were collected by the cowkeepers did not include any flowering ones. In addition, the cowkeepers carefully selected the cane tops and discarded the very mature ones. From those selected, the outer older leaves and all dry ones were discarded. Some cowkeepers had the habit of chopping off and discarding about a third of the leaf blade at the tapering end when they were collecting their bundle of cane tops in the field.

The cows' eating behaviour, as noted during the farm visits, was confirmed in numerous interviews with the cowkeepers. The cows consistently started to consume the cane tops at the sheath bundle end. They consumed the whole of the sheath bundle portion first, then started to eat the leaf blade (Figure 1). They consumed only part of the leaf blades on both sides of the midrib. It was estimated that from one third to one half of the leaf blade was thus consumed.

It was observed in the wet uplands that *Ischaemum aristatum* (herbe d'argent) generally formed the bulk of the fodder that was given to cows during the sugar cane inter-crop season whereas other assorted forages, singly or mixed, were given in small amounts of a few kilograms per cow. In the drier area, *Plantago lanceolata*, *Digitaria didactyla*, *Stenotaphrum dimidiatum* and young cane regrowths (about 50 cm high) generally formed the bulk of the fodder in the inter-crop season. A few kilograms of other assorted forages, singly or mixed, were also given to the cows.

Figure 1. Sheath bundle and leaf blade fraction of cane top.

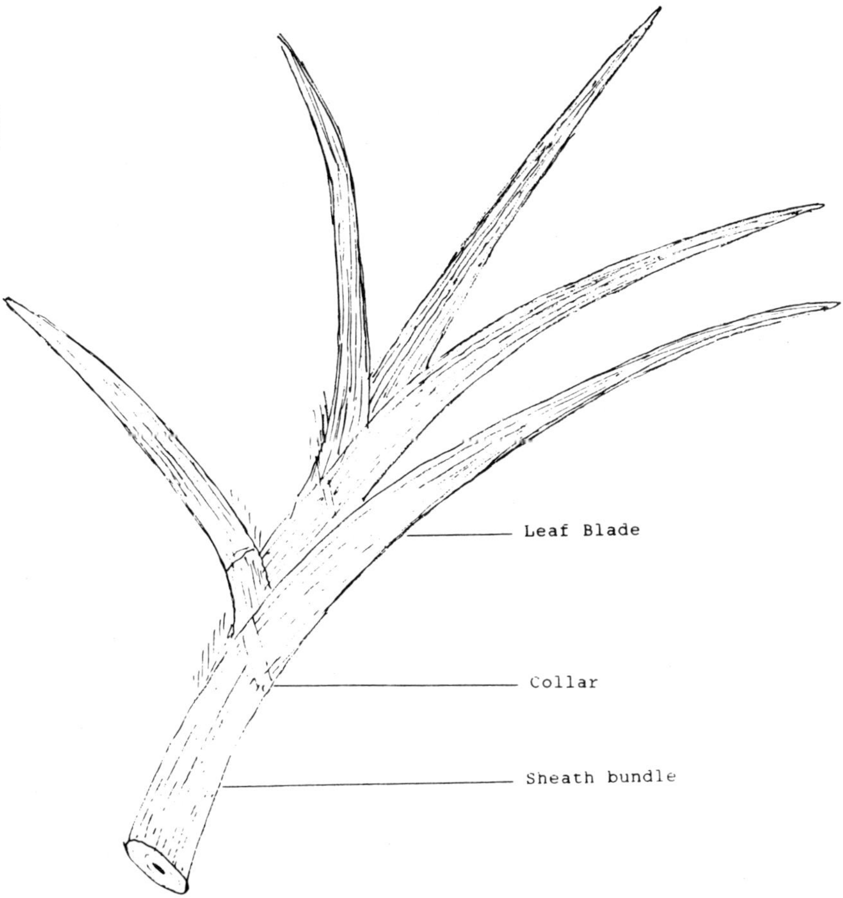

Chemical composition

The sheath bundle fraction had a lower dry matter and crude protein content (19.3% and 5.5% respectively) than the leaf blade (31.3 and 7.3%) or the assorted forages (23.1 and 9.7% respectively). The crude protein content of the assorted forages and crop residues (9.7%) was quite interesting. The leguminous forages contained up to 14% crude protein (Table 3).

Table 3. Chemical composition of the sheath bundle and leaf blade fractions of cane top, the assorted forages and crop residues (on dry matter basis) (Boodoo et al., 1988b).

	Dry Matter			Crude Protein			Crude Fibre		
	(n)	Mean	SE	(n)	Mean	SE	(n)	Mean	SE
1. Leaf blade*	(23)	31.3	0.9	(25)	7.3	0.4	(22)	32.9	0.8
2. Sheath bundle*	(23)	19.3	1.0	(24)	5.5	0.3	(22)	31.9	0.5
3. Assorted forages and crop residues	(49)	23.1	1.0	(49)	9.7	0.5	(15)	31.0	1.2
Range for assorted forages etc.									
Min.		13.3			5.0			14.5	
Max.		40.9			14.6			37.2	

* These fractions were prepared from one original big bundle (about 7 kg) of cane tops.

Rate of degradation in nylon bag

Results of the nylon bag study of the various types of forages are shown in Table 4. These data show that the sheath bundle fraction degraded much faster than the leaf blade at all the time intervals studied. On the other hand, the sheath bundle and assorted forages show remarkably similar DM degradabilities from the 16, 24, 48 and 72 hour incubations. The rate and extent of degradation supports the view that the sheath bundle and assorted forages are good quality roughages in this environment. Their potential degradability is approximately 26% higher than that of the leaf blade.

Table 4. Mean dry matter degradability values of the cane top fractions and assorted forages at various intervals in nylon bags (Boodoo et al., 1988) (SE in brackets).

	n	Incubation Time (h) 0	16	24	48	72
Leaf blade	18	12.0 (0.6)	27.0 (1.1)	34.3 (1.0)	47.0 (1.1)	53.4 (1.1)
Sheath bundle	18	21.5 (0.7)	46.7 (1.5)	55.5 (1.3)	62.6 (1.1)	67.2 (0.8)
Assorted forages	18[1]	16.4 (0.7)	44.5 (1.6)	54.6 (5.2)	63.5 (1.3)	66.4 (1.3)
LSD (5%)		2.2	5.3	4.7	4.2	3.8

[1]These 18 samples were a random selection from 49 original ones.

Intake

Table 5 summarizes the data on forage consumption, liveweight and DM intake.

Table 5. Quantity of forage offered and eaten. Liveweight of the lactating cows and their DM intake on diets consisting predominantly of (a) cane tops and (b) mixed grasses with crop residues (Boodoo et al., 1988b). (SE in brackets).

	Wet uplands		Dry Area	
	Cane tops	Mixed grasses	Cane tops	Mixed grasses
n	7	7	9	5
Fodder offered (kg fresh/day)	69.3 (4.11)	68.0 (4.3)	73.1 (1.0)	62.6 (4.2)
Fodder eaten (kg fresh/day)	43.6 (3.5)	57.2 (2.2)	52.1 (1.8)	49.2 (2.2)
Surplus feed (% of amount eaten)	58.9%	18.9%	40.3%	27.2%
Liveweight of cows (kg)	342 (21.8)	353 (13.3)	364 (11.4)	311 (16.0)
Fodder intake (kg DM/day)	11.8 (0.9)	13.5 (0.7)	14.1 (0.5)	11.0 (0.5)
Concentrate intake (kg DM/day)	2.7 (0.3)	3.3 (0.4)	3.5 (0.4)	2.9 (0.5)

The cows were offered forages _ad libitum_ such that the surplus amounted to 19-58% of the amount eaten, especially sugar cane tops which are abundant during the harvest season. This surplus gave them plenty of scope to select the best parts and this resulted in a high intake of fodder dry matter (about 3.7% of body weight).

DISCUSSION AND CONCLUSION

Cottonseed cake provided slightly more protein than cowfeed at the levels used. However, cowfeed, because of its molasses content would have provided more energy than cottonseed cake. It is therefore interesting to note that similar levels of milk were obtained with these two supplements, which point to protein being the primary nutrient limiting milk production under these conditions.

The traditional practice of selectively feeding a variety of forages has sound scientific basis. With time, the cowkeepers have developed an appreciation of forage quality. The present data show that the rate and extent of degradability of the sheath bundle fraction of cane tops and the assorted forages are the basis for good milk production in the villages, compared to the larger systems of production where there is less attention to forage selection.

The local Creole breed has sufficient potential for milk production, given the climate and the resources available on village smallholdings in Mauritius. There would therefore appear to be no advantage to be gained from the importation of exotic breeds for milk production.

REFERENCES

Bennie, J.S.S. 1956 The Mauritius Creole breed of Milch Cow. _Empire Journal of Experimental Agriculture._ 24 (95).

Boodoo, A.A., Ramjee, R., Hulman, B., Dolberg, F. and Rowe, J.B. 1988a The effect of supplements of cowfeed and cottonseed cake on milk production in Mauritian villages. Paper presented at seminar Milk and Beef Production in Mauritius, Division of Animal production, Ministry of Agriculture, Fisheries and Natural Resources. June 7 -8, 1988.

Boodoo, A.A., Ramjee, R., Hulman, B., Dolberg, F. and Rowe, J.B. 1988b Evaluation of the basal forage diet of village cows. In: Milk and Beef Production in Mauritius, _loc. cit._

Boodoo, A.A., Ramjee, R., Hulman, B., Dolberg, F. and Rowe, J.B. 1989 The response of Creole, Friesian and Friesian cross cows to concentrate supplementation on village smallholdings in Mauritius. Animal Production. In press.

FAO. 1971 Milk and Meat Project. FAO, Mauritius. Working reports Nos. 1, 5 and 17.

Gaya, H., Hulman, B. and Preston, T.R. 1982 The value for milk production of different feed supplements: Effect of cereal protein concentrate, poultry litter and oilseed meal. Tropical Animal Production 7: 134-137.

Ma Poon, L.K., Delaitre, J.C. and Preston T.R. 1977 The value for milk production of supplements of mixtures of final molasses, bagasse pith and urea, with and without combinations of maize and groundnut cake. Tropical Animal Production 2: 148-150.

TRAINING IN THE DEVELOPMENT OF FEED RESOURCES

by

R.W. Froemert

INTRODUCTION

The FAO Regional Dairy Development Training Team for Asia and the Pacific has been engaged since 1972 in training of extension workers and milk plant personnel. Some 2500 officers from 15 countries of the Asia and Pacific region have been participating in specialized courses of 2 - 6 weeks duration and self-teaching programmes. Training of smallholder milk producers consists of short courses and demonstrations at dairy development training units (small demonstration farms) in Sri Lanka, Thailand and Indonesia and the operation of mobile dairy extension units in selected areas of Thailand and Indonesia.

Having noted the lack of specialized knowledge and experience of field extension workers in forage preservation, the Team designed training activities in making silage and hay, which provide the trainee extension worker with adequate practical experience to handle these subjects competently on the village level. The Team has contributed to the development of small scale preservation practices: silos, hay drying racks, hay stores and feeders, which fit the means, skills and requirements of the small farmer. A better utilization of various field crop residues and industrial by-products as cattle feed has been promoted.

The successful introduction of new practices benefitting the village community depends a great deal on the proper presentation to the farmer. Part of all training activities in forage production/ preservation and use of alternative feed resources is the development of training material, "persuasive" posters and handouts for the farmer.

DEVELOPMENT WITH AND FOR THE SMALLHOLDER

Sites and Modes of Training

A sound development of smallholder (dairy) farming and the designing of realistic extension and training programmes supporting this development require, first and above all, to listen to the farmer. The small farmer is a professional. His profound knowledge of his environment, his experience and practical skills are valuable assets of development.

The researcher (extension worker) must realize that it is no use to drop a new idea (farming practice) on the practical man and hope for the right adjustments at the practical level. Developing his ideas requires the researcher (extension worker) to take into account the height of the hill to be reached, the obstacles in its way and the weight of the idea or recommendation.

Introducing to smallholder farmers a new method of forage utilization, such as silage or hay making, requires considerable experience on the part of the extension worker. A thorough analysis of training courses, conducted in 7 countries of the region for extension workers in dairy husbandry, reveals that the time allocation for practical forage preservation is completely inadequate. The lack of experience endangers the success of extension activities on the village level.

To reinforce practical training facilities for trainee extension workers and smallholder milk producers, the RDDTTAP (Team) has set up small scale demonstration farms (Dairy Development Training Units) in 3 countries of the region: Sri Lanka, Thailand and Indonesia. It is here that small numbers of farmers gather for repeated one-day training activities and are acquainted with new farming practices and developments. They include aspects of forage production/preservation.

Besides training rendered by extension workers operating from DDTUs, training is being extended to other areas relevant for dairy development through mobile extension units (Thailand and Indonesia). The Team developed and is developing a special tool for mobile operations: the training kit. It contains a comprehensive collection of equipment and training material required by the extension worker to demonstrate particular points of the subject matter and to enhance the understanding of it by the farmer. Training kits on forage production and preservation are in the state of being completed.

An integration of DDTUs, mobile dairy extension units and agricultural colleges in support of the resident extension worker is envisaged as part of the development strategy pursued by the Team.

On the basis of encouraging experiences in South America with self-teaching programmes, the Team has introduced self-teaching programmes on ration formulation in Sri Lanka. The programme is aiming at a better knowledge of the field extension worker on ruminant nutrition and balanced feeding, as well as to encourage the establishment of a closer network for his support by research and teaching institutions.

SILAGE MAKING

A Technology for the Smallholder

If silage making and feeding is to be introduced to smallholders, the extension worker tends to organize a "Field Day" on one of the nearby government farms. What does the small farmer observe? An impressive technology - but nothing which could fit his own farming conditions. The extension worker, with the tower silo in his mind, may even build one on his own. It is too big for one farmer, so he calls it "community silo." But is it practical? A more simple and inexpensive approach to "persuade" smallholders to adopt silage making is still the famous pit silo. But what about silage quality? What about losses reducing the amount of feedable plant matter?

Silo pits are of temporary nature. When not reinforced (costs!), they may even encourage erosion and their walls collapse during the rain. If silage making becomes a regular practice on a smallholding, other building materials may be considered, e.g. bamboo.
Silos must be airtight to provide anaerobic conditions for fermentation. Wire netting (chicken-wire) can be connected to the inner wall of the silo, to provide the reinforcement for a thin layer of concrete. A plastic jacket may provide for the same conditions, but lasts only through one silage campaign and feeding period.
Even after years, when the bamboo structure may have disintegrated, the concrete wall is still intact, the silo still usable.

The time which is required to fill a silo and seal it appears to be the most decisive factor for the production of quality silage. Chopping of forage is a necessity for appropriate compaction of the forage mass. But chopping by hand needs many hands and is time consuming.

Forage choppers of many kind are available. They often exceed the capacity which is required for the smallholder's operation: chopping of forage for fresh feeding <u>and</u> ensiling. Animal driven forage choppers have been successfully used for many years in India and Pakistan. The two-wheeler tractor is a common piece of equipment for the small farmer in Thailand as elsewhere in the region. It is used for all the land tilling operations connected with the cultivation of paddy. A technology, which is not to burden the smallholder and to increase his cost of production, must be linked to existing elements and resources. Two-wheeler tractors can be engaged as power source in forage chopping (e.g. in Sri Lanka).

<u>Training and demonstration methods which can persuade the smallholder</u>

Methods and practices used in training of smallholders must help the trainee to identify the demonstrated practices and techniques as within his reach and means. To introduce silage making on smallholdings through the use of big machinery and equipment is a didactic mistake. Sometimes also too many trainees are gathered, too many trainees not occupied. The demonstration, more often than not, has no follow-up. The farmer observes that grasses can be "buried". Since he does not see the result, he may think: 'What a waste!'.

Figure 1a.

TOPIC : INTRODUCING SILAGE MAKING
THONGSRI AND HIS ANIMALS HAVE A WONDERFUL TIME
THONGSRI RESTS WHILE HE SHOULD PREPARE HIMSELF FOR THE DRY SEASON

Figure 1b.

TOPIC : INTRODUCING SILAGE MAKING
MANIT HAS NO PROBLEM TO FEED HIS ANIMALS DURING THE DRY SEASON. HE PRESERVED GRASSES.
MANIT EXPLAINS WHAT SILAGE IS AND WHAT IT NEEDS TO MAKE SILAGE
(poster: RDDTTAP, 1988)

The chart (Figure 1) demonstrates the need for a well timed and organized training activity on small scale silage making, during which all steps from silo construction to silage feeding are shown. Trainee extension workers work together and learn together with the farmer during a field day on a smallholding.

Prior to the introduction of silage making, farmers are taught the basic requirements for the production of quality silage (chopping, quick filling, compacting, sealing). Chopping by hand can be employed for handling smaller quantities of forage to produce sample silage. Well cleaned, 44-gallon oil drums furnished with a plastic jacket may be used as silos. In connection with the Team's work at DDTUs linked to agricultural colleges, concrete sewage pipes (height 1 m, diameter 1 m) have been used as silos.

With a sufficient number of experimental silos, the impact of various processing practices of forages on the quality of silage can be demonstrated (e.g. unchopped = insufficient compaction = continued respiration of plant matter = suppression of lactic acid producing bacteria). The proper closing (sealing) of the silo immediately after completing the fill is demonstrated.

Silos need to be identified with regard to the plant matter ensiled, its quantity and the treatment applied. Brief descriptions, on boards of the silo wall, also make it easier to explain new developments and practices to the visitor to the demonstration site.

The circular silo erected from bamboo poles (height 1.80 m, diameter 2 m) can be filled to contain around 2 tons of forage. A compaction of 300 - 350 kg/m^3 by treading is only possible if the forage is chopped.

Trainee extension workers and farmers are made acquainted with the criteria for the wanted fermentation. On the basis of organoleptic tests, they produce a quality score. The decisive test is the animal's intake. Silage of good quality has been produced from sugar cane tops in an experimental silo. Tops were chopped, well compacted and 4% molasses added.

Forage planning and the use of crop residues and industrial by-products

The eager extension worker may wish to do something immediately about forage preservation in his working area. But has he looked carefully into the requirements and conditions for silage making? Above all, has he looked into the farmers' "feeding calendar" to see if silage is really needed and if there is sufficient surplus of forage to make silage preservation feasible?

Once it has been determined that silage needs to be produced, the quantity of forage required has to be specified and the silo capacity and dimensions be calculated. Trainee extension workers have to be engaged in exercises along these lines. An overestimated forage quantity leads to an only partly filled silo, increasing the surface area for the fodder mass and does often result in inferior silage quality.

Participants of Team-supported training courses are requested to establish an inventory of available feed resources. If there is no surplus of forage crops, field crop residues and industrial by-products can fill the gap. Sugar cane tops have been recognized as a potential feed resource both for being fed fresh and as silage. The vast quantities of sugar cane tops available during the campaign period suggest that they are wasting feed resources by improper feeding practices. Farmers are advised to construct inexpensive feeders (bamboo feeders at DDTU Nakhon Sawan) and to chop cane tops to reduce waste. The Team has encouraged extension workers to make use of pineapple cannery waste as cattle feed. Some years ago, pineapple waste was discarded at dumping places, creating problems of pollution. A number of training activities were conducted to demonstrate silage making from pineapple cannery waste together with paddy straw.

HAYMAKING

"The growing season of grasses coincides with heavy rain. Haymaking is not possible." This statement may justify the common observations made during visits to small and big dairy farms alike in this region that no hay can be found in calf pens. But if so, there is some colourless, stale dry roughage. It deserves to be called straw - but not hay.

Based on hay drying practices experienced earlier in temperate climates, the Team, during training courses for extension workers, attempted to develop a technology of hay curing suitable for the smallholder and applicable during the rains.

The "classical" tripod worked well in Pakistan, (forage crop: *Trifolium alexandrinum*), but was found unsuitable for the conditions to be encountered during haymaking in countries like Thailand, Sri Lanka or Indonesia. Layers of wilted forage on tripods become too compacted, thus hindering air circulation and causing the development of moulds.

Satisfactory results were achieved with a drying fence. It is constructed by inserting (4) poles in a particular pattern in holes driven (1.10 m apart) into the ground using crowbars. Then layers of shortly wilted forage are placed on ropes connecting the poles. The forage on fences is covered between rains by mats made from palm leaves.

A drying rack called the "Hurdle" proved to be the most easy to construct and practical means for drying small quantities of hay efficiently during the rains. Hurdles, made from timber or bamboo, can be used in flat as in hilly areas. After a few hours of pre-drying on the ground, encouraged by continuous turning, the forage is packed onto the horizontal runners. For the first or second day the roof may be removed during sunny spells to encourage fast drying. Thereafter, the hood is kept until transfer of the hay to the site of feeding. Permanent shading prevents bleaching (i.e. loss of carotene).

Research workers at the National Dairy Training and Applied Research Institute (NDTARI) in Thailand, supported by the Team, developed an extremely successful drying technique using 1-pole racks ("Heinze') as the basic element. The drying rack consists of poles, a dark plastic sheet in the bottom of the drying rack which collects the suns rays, and transparent plastic sheeting provides protection for the forage mass from the rain. Smaller drying racks of this kind have been developed and are regularly demonstrated by NDTARI to participants of training courses. The drying process takes 1-3 days and produces a green coloured, aromatic hay.

Together with hay curing practices suitable for the small farmer, baling of hay by "baling box" was introduced, particularly to avoid losses through leaf shattering (fragmentation) during transport.

Hay may be stored in many ways. Since the smallholder is encouraged to produce small quantities of hay for calves as often as possible to avoid spoilage occurring during extended storage periods, only small quantities need to be kept. A combination of a roofed silage pit and a hay drying/ storage platform makes excellent use of limited place on smallholdings (e.g. in Sri Lanka). Hay may also be stored on a well-aired platform established on top of the area, where calves are housed (e.g. DDTU Nikaweratiya, Sri Lanka).

In search of practical, low cost methods for handling, storing and feeding of hay, the Team has utilized the farmer's skills for wickerwork. Hay for a number of calves can be stored in roofed self-feeders which are best placed in exercise yards.

Problems in curing quality hay during the growing season of grasses, which coincides with heavy rain, can be mastered by selecting appropriate drying practices and sticking to a flexible management which makes maximum use of dry and sunny spells. The remaining problem is the storing of hay under condition characterized by a high humidity.

MAKING BETTER USE OF PADDY STRAW THAN BURNING

Cereal straw of different kind forms the bulk of crop residue available immediately after threshing. Draught animals are often fed entirely on straw without losing body weight. Though low in feed value, straw has to be used extensively on many smallholdings to form a major part of the maintenance ration for dairy animals. But straw is also burned, year by year, after harvest and threshing of grain crops.

The Team has encouraged, through training of extension workers and farmers, to better utilize available straws for cattle feeding, either by ensiling it together with low dry matter industrial by-products (cannery waste) or by ammoniation. For the treatment of cereal straws (mainly paddy), the Team promoted, as a temporary arrangement without investment in permanent structures, the covering of treated straw stacks by plastic sheets (6% urea, 3 weeks storage).

Similar to training activities in silage making, demonstrations to farmers on straw treatment have to include all stages from processing to feeding. Trainee extension workers, at the end of the course, develop their own training aids and conduct a field day at a farm site.

Lately, it is being tried in Pakistan to combine the threshing of cereal crops and the treatment of straw with urea, with the aim of making better use of available manpower at the time of threshing.

THE NEED FOR MINERALS AND SMALL-SCALE BLOCK MAKING

The making of mineral blocks was originally introduced as a training activity for extension workers and farmers as a didactic means to make extension workers and farmers aware of the importance of minerals in support of growth, milk production, breeding efficiency and reproductive performance. To talk to farmers about the need of animals for calcium, phosphorus, potassium, etc. remains highly theoretical and is generally beyond his comprehension. The presentation, during a demonstration to farmers, of chemical compounds containing the required elements, the weighing (dosing) of compounds and their mixing according to an explained formula, brings them nearer to understanding the complex nature of this feed ingredient.

In following up on this aspect of training, more efficient presses were developed, which have made it possible to produce mineral blocks in bigger quantities for sale to dairy farmers (e.g. in Thailand). Formulae based on observed mineral imbalances in cattle rations have been developed.

[illegible] PUBLICATIONS PUBLISHED BY [illegible]

1. Agricultural Biotechnology in the Developing World — Rs. [illegible]
2. Aquaculture & Risk Management / An FAO Fishing Manual / FAO [illegible] & C.E. Nash (PB) Rs. [illegible]
3. Aquaculture Project Formulation, An FAO Fishing Manual / [illegible] — Rs. [illegible]
4. Aquaculture Systems & Practices / A [illegible] Review. An FAO Fishing Manual / [illegible] — Rs. [illegible]
5. Climate Change, Forests & Forest Management / An Overview / [illegible] — Rs. 205
6. [illegible] Aquaculture — Rs. [illegible]
7. [illegible] the Tropics / [illegible] — Rs. [illegible]
8. [illegible] Manual for the [illegible] / [illegible] — Rs. [illegible]
9. Guidelines for the Promotion of [illegible] Aquaculture Development / I.C. [illegible] — Rs. [illegible]
10. Plant Tissue Culture: An Alternative for Production of Useful Metabolites — Rs. [illegible]
11. [illegible] Structures & Layouts—Simple Methods [illegible] — Rs. [illegible]
12. [illegible] Vaccines in Developing Countries / [illegible] — Rs. 1200
13. Quality Declared Seed — Rs. [illegible]
14. [illegible] S.B. Mathur & H.K. Manandhar — Rs. 380
15. Reservoir Fisheries of India — Rs. 1290
16. Small-Scale Poultry Processing / [illegible] — Rs. 395
17. Sustainable Animal Production from Small Farm Systems in South-East Asia / [illegible] — Rs. 295
18. Tropical Animal Feeding: A Manual for Research Workers / Preston [illegible] — Rs. [illegible]
19. Veterinary [illegible] — Rs. [illegible]